DEEP
LEARNING
FOR PHYSICS
RESEARCH

DEEP
LEARNING
FOR PHYSICS
RESEARCH

Martin Erdmann
RWTH Aachen University, Germany

Jonas Glombitza
RWTH Aachen University, Germany

Gregor Kasieczka
University of Hamburg, Germany

Uwe Klemradt
RWTH Aachen University, Germany

World Scientific

NEW JERSEY · LONDON · SINGAPORE · BEIJING · SHANGHAI · HONG KONG · TAIPEI · CHENNAI · TOKYO

Published by

World Scientific Publishing Co. Pte. Ltd.

5 Toh Tuck Link, Singapore 596224

USA office: 27 Warren Street, Suite 401-402, Hackensack, NJ 07601

UK office: 57 Shelton Street, Covent Garden, London WC2H 9HE

Library of Congress Cataloging-in-Publication Data

Names: Erdmann, Martin, 1960 February 6– author.

Title: Deep learning for physics research / Martin Erdmann, RWTH Aachen University, Germany,
 Jonas Glombitza, RWTH Aachen University, Germany, Gregor Kasieczka, University of
 Hamburg, Germany, Uwe Klemradt, RWTH Aachen University, Germany.

Description: New Jersey : World Scientific, [2021] | Includes bibliographical references and index.

Identifiers: LCCN 2021024472 (print) | LCCN 2021024473 (ebook) |
 ISBN 9789811237454 (hardcover) | ISBN 9789811237461 (ebook) |
 ISBN 9789811237478 (mobi)

Subjects: LCSH: Physics--Data processing. | Physics--Research. | Machine learning.

Classification: LCC QC52 .E73 2021 (print) | LCC QC52 (ebook) | DDC 530.0285--dc23

LC record available at https://lccn.loc.gov/2021024472

LC ebook record available at https://lccn.loc.gov/2021024473

British Library Cataloguing-in-Publication Data

A catalogue record for this book is available from the British Library.

For any available supplementary material, please visit
https://www.worldscientific.com/worldscibooks/10.1142/12294#t=suppl

Preface

Deep learning has become an electrifying term thanks to numerous successes, for instance, in pattern recognition, generative modeling, and even playing games. In recent years, basic deep learning techniques have been established, adapted, and further developed in physics to such an extent that conveying its basic concepts via a textbook aimed at physicists is needed. Deep learning is a rapidly developing field, but we expect this work to form a solid foundation for integrating new developments in the future.

The textbook is based on lecture series and exercises that we have been offering for several years to undergraduate and graduate students in physics. In our own research, machine learning methods are a means to an end of gaining deeper insights into questions in physics. Therefore, this textbook reflects the perspective of research applications of deep learning in physics. Our goal is to lower the threshold for getting started with deep learning, describe what is behind the commonly used professional vocabulary, and build up a comprehensive understanding for practical use.

Some of our students look for further theoretical insights and understanding the foundations of machine learning in computer science and mathematics courses. There are several computer science textbooks on this subject, where we ourselves like to refer to the comprehensive compilation of Ref. [1]. For new developments of deep learning in physics, refer to respective reviews where, e.g., in the sub-field of particle physics, informative access is provided in Ref. [2]. A book with focus on theoretical physics and deep learning is provided by Ref. [3].

For our textbook on deep learning, only little prior knowledge is needed. Required are mathematical and physical knowledge from the first semesters of the bachelor's degree in physics and some familiarity with Python as the programming language used for the exercises. Each chapter contains the learning objective at the beginning and a summary of key points at the end. Boxes visually highlight the description of examples and experiments within the chapters. Most chapters also include hands-on exercises, e.g., implementing networks for a specific task. Additional material for the exercises, such as pre-structured Python scripts, data, and sample solutions can be obtained from `www.deeplearningphysics.org`.

For the implementation of the exercises, we recommend installing freely available software packages (Keras, Python) on your own computing resources. Usually, the CPU of a standard laptop computer is sufficient for training small networks. Some notebooks also allow using the GPU for training. Alternatively, commercial offers can be used, which are available at low cost or free of charge for academic or moderately-scaled applications. For the implementation of lectures or courses based on this book, we recommend setting up central computing resources, e.g., via JupyterHub or similar systems.

On the one hand, we hope that our textbook provides a structured insight into the foundations, possibilities, and limitations of deep learning. Second, we wish the readers to successfully advance their own research using deep learning and its data-driven insights. We would be grateful for any information on errors in the textbook to `authors@deeplearningphysics.org`.

We thank our colleagues for conversations at international big data and machine learning workshops, and especially our co-authors of research publications for in-depth discussions. Many arguments were sharpened by discussion with students in our machine learning courses in Aachen and Hamburg. We are grateful to many colleagues of our departments who supported this book and the deep-learning initiative in various aspects, and to Dr. D. Walz and Dr. S. Schipmann for substantial help in preparing the initial lectures at RWTH Aachen University. Furthermore, we would like to thank the team at World Scientific Publishing for their excellent support in publishing this textbook. Finally, we acknowledge the helpful machine-learning based platforms deepL (`www.deepl.com`) and grammarly (`www.grammarly.com`).

Prof. Dr. Martin Erdmann, Jonas Glombitza,
Jun.-Prof. Dr. Gregor Kasieczka, Prof. Dr. Uwe Klemradt
Aachen and Hamburg in Germany, April 2021

Contents

Standard Architectures of Deep Networks   67

Deep Learning Advanced Concepts   215

PART 1

Deep Learning Basics

In this part, we will explain the elementary components of neural networks and their interplay. We will show how the network parameters can be optimized and how the quality of this optimization process can be controlled.

Chapter 1

Scope of this textbook

Scope 1.1. Research questions to be answered from measurement data require algorithms. A substantial part of physics research relies on algorithms that scientists develop as static program codes. An alternative is to assemble program modules with adjustable elements that are autonomously tuned in a learning process. This method is called machine learning and has recently gained increasing attention in physics research. An ultimate goal in mastering such machine learning methods is data-driven knowledge discovery.

In this chapter, we classify machine learning methods for data analysis in physics. We note the new challenges for physicists when using machine learning compared to previous data analysis techniques and remind of the unchanged task of obtaining both the results and their uncertainties. We relate to current developments in computer science and artificial intelligence. Finally, we explain the structure of this textbook and encourage active participation in hands-on exercises.

1.1 Data-driven methods

The development of new technologies is decisive for advances in fundamental and applied research. Particular attention is paid in this textbook to advancements in modern data analysis methods: they are the key to gaining new insights from measurement data and achieving considerable added value from an experimental apparatus. This leads to more precise scientific results, accelerates processes of knowledge acquisition, or even enables discoveries.

A transformative process is currently observed through so-called data-driven methods in the context of machine learning, especially deep learning. Here, machines are directed by physicists to autonomously build algorithms and extract additional information from the data. In a number of examples, the machines' results appear to outperform results previously obtained with physicists' algorithms.

1.2 Physics data analysis

To answer a scientific question from measurement data, a concrete algorithm is usually formulated, programmed, and executed on the data. In the physics beginners' lectures and laboratory courses, experiments are carried out, which have only a small number of observables such as time, length, voltage, current readings, etc. For evaluating such experimental data, physics-motivated algorithms are developed by hand. In Fig. 1.1, this procedure is referred to as a rule-based system.

The recorded data is quite different when investigating, e.g. samples of innovative materials at modern synchrotron radiation laboratories. A typical example is two-dimensional scattering data with a million pixels resulting from *in-situ* experiments, recorded at high image rates, which need to be analyzed according to a scientific question. The challenges

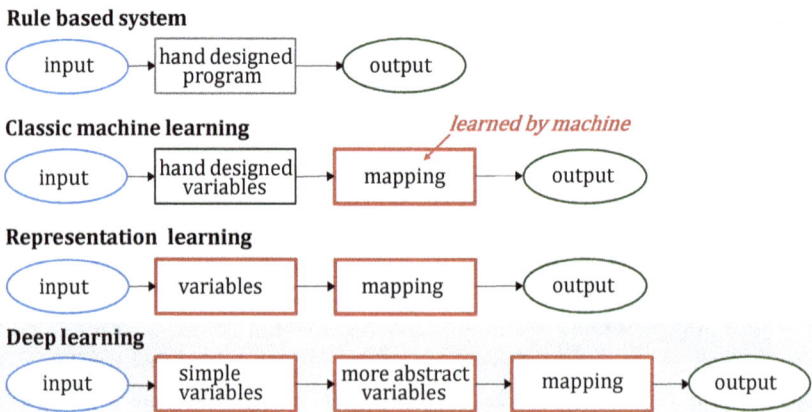

Fig. 1.1. Different levels of data analysis methods, from hand-designed programs via classical machine learning to representation learning and deep learning.

include filtering relevant information from the images and developing suitable algorithms to decipher the data's physics messages.

Another example is large-scale particle collision experiments. Therein millions of sensors are hit simultaneously, and their data are recorded. For each collision, the measurement data are processed with algorithms to reconstruct particle trajectories or to find clusters of particles, thus reducing the number of observables to a few thousand and storing them for further analysis.

Since the formulation of a suitable algorithm for answering scientific questions using so many observables proves to be extremely difficult, researchers have worked with machine learning methods for many years.

1.3 Machine learning methods

A definition of machine learning, following Arthur Samuel, is (1959) [4]:

> *Machine learning is the subfield of computer science that gives computers the ability to learn without being explicitly programmed.*

A common example of a classical machine learning method is the *decision tree* (Fig. 1.2). A choice of physics-motivated observables x_1, x_2, x_3 are transferred to the decision tree, which has adjustable parameters to perform data selections, subsequently on the given observables starting at the top of the figure. Using a set of data for learning, the order of observables and the selection criteria are automatically optimized.

From the input values, a target variable is formed, with which a scientific question is answered. The figure shows the classification of objects as belonging to class signal S or class background B. Each box indicates a decision on the object path, and the circles represent the vote *signal* or *background*. The numbers are for demonstration purposes only and reflect probabilities of correct and false decisions.

In Fig. 1.1, the decision tree method is referred to as 'mapping'. It is classified as 'classic machine learning' as the observables provided to and used by the decision tree are those selected by physicists.

The next more advanced level in Fig. 1.1 is referred to as 'representation learning'. Here observables are given to a machine directly without a physics-motivated pre-selection or an intermediate step of combining observables. Instead, the machine learning method is supposed to perform this work before the mapping process. An exemplary challenge consists of combining location and time data of an object for obtaining its velocity.

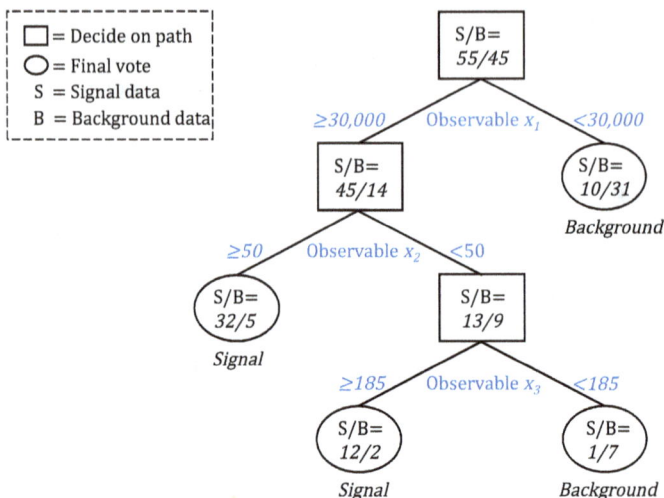

Fig. 1.2. The decision tree is a machine learning method predicting object assignments to classes S or B based on selections using observables x_1, x_2, x_3.

1.4 Deep learning

Deep learning methods are on everyone's lips thanks to the impressive examples of image and speech recognition and the development of self-driving cars. These methods employ so-called *deep neural networks* where a simplified sketch can be found in Fig. 1.3.

Read the network from the left input x_1, x_2, x_3 to the right output value z. The network is structured here in two inner *layers*, each consisting of three *nodes* located one above each other (circular symbols). The inner layers are often referred to as *hidden layers*.

The network parameters have already been adjusted here, resulting in characteristic paths through the network. The blue connections symbolize the fixed network parameters, which combine the inputs at subsequent nodes. The red network nodes show the strength of the node response based on the input values. The input and output values, the coloring of the weight parameters, and the node responses are only for illustration purposes.

The three input values x_1, x_2, x_3 determine which paths contribute how much to a calculation by the network. This results in a mapping of the input values to the output value z. We will present here the basic strategy

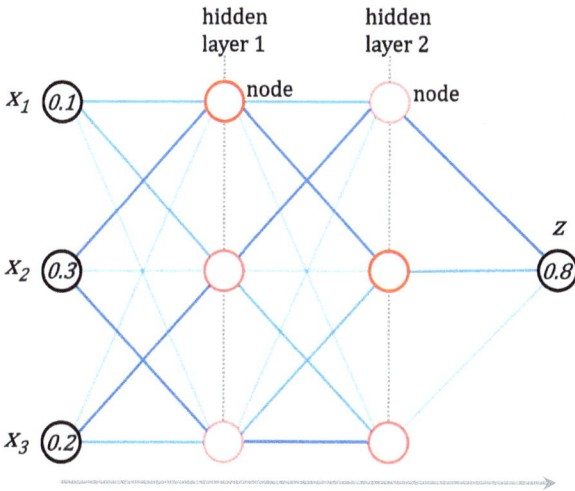

Fig. 1.3. Neural network combines observables x_1, x_2, x_3 in several stages before predicting an output value z. The two inner network *layers* (*hidden layers*) have three *nodes* each. The blue links symbolize the weights of the connections, the red circles the strength of the response based on the input values. Input values, output value as well as coloring are for illustration purposes only.

only but will, of course, explain neural networks in great detail later in this book.

With the application of deep neural networks in physics, a decisive additional step is made beyond representation learning in Fig. 1.1. From all observables provided to the network, numerous new combinations of observables are formed in the network, from which again numerous further combinations are generated, etc. This property is what one of the founders of deep learning, Yann LeCun, then calls a *deep* neural network [5]:

> It's deep if it has more than one stage of non-linear feature transformation.

Translate the word *feature* commonly used in computer science with *observables* or *variables* used in physics. The millions of adjustable parameters are used to decide what these multiple observable combinations will look like.

With the help of the measurement data themselves — alternatively, with simulations of the measurement data — the millions of network parameters are adjusted so that optimal observables are created for answering the scientific question. The driver for adjusting parameters is the so-called objective function which provides some distance measure between the network prediction and an ideal result. This distance measure is minimized during parameter optimization.

It is important to understand that it is not physicists who specify the best-suited physics observables. The deep neural network *autonomously* forms the optimal observables with the input measurement data and an objective function. Thus, the ultimate goal is the development of machines that autonomously derive new findings from measurement data.

To answer a scientific question, the algorithm to be applied in analyzing the measured data consists of the network itself with the optimally adjusted parameters. Accordingly, the physicists' work focus has shifted from finding suitable observables and developing algorithms towards designing suitable network architectures and finding appropriate objective functions for setting their millions of parameters. These new tasks are the focus of this textbook.

1.5   Statistical and systematic uncertainties

For an experimental measurement result in physics, estimates of *statistical* and *systematic uncertainties* are mandatory. The procedure for data analyses with neural networks is no different from usual analysis procedures.

Neural networks produce a mapping from n input variables to m output values. Statistical uncertainties, which are, e.g. based on the number of events or observations, are obtained as usual via Poisson statistics. Systematic uncertainties, such as calibration inaccuracies, can be investigated as usual. For example, a change in a measurement device's calibration will change the network's input data. The mapping properties of the network will reflect this calibration change in the network output correspondingly.

In Part 3 of this textbook, we will explain which uncertainties arise from adjusting the network parameters. We will also show methods to get insight into the functioning of the network. In Part 4, we will further discuss solutions for possible problems arising in adjusting network parameters with data from simulations that do not exactly correspond to the measured data.

1.6   Causes for rapid advancements

The rapidly accelerated development of deep neural networks has various causes. The progress is mainly based on four developments in computer science that have led to impressive successes in the field of image and speech recognition:

- Various architectural concepts for networks have been developed that are tailored to the structure of the problem to be solved. Simultaneously, technologies such as improved optimization methods for the parameters of deep neural networks have been developed.
- Access to Graphics Processing Units (GPUs) allowed the optimization of network parameters to be carried out much more time-effectively than before. As a result, many more connections in the network can be realized, and more complex decisions can be made, e.g. a larger number of categories in the classification of objects.
- Large data sets for the optimization of networks have been collected, e.g. by the millions of images with classifications (animal species, objects of all kinds, ...) collected on the worldwide web.
- User-friendly *software libraries* have been developed, some of which are made available as open-source programs. These libraries enable people with programming skills to efficiently design networks.

Machine learning counts as a subcategory of *artificial intelligence* (AI), as is *representation learning* and *deep learning* (Fig. 1.4). We emphasize the machine learning aspect as physicists play a decisive role in designing and executing the machines for successful scientific results. We cite here the AI researcher Francois Chollet with the statement:

> *In AI, system should be understood as including the human engineers. Most of the data-generalization conversion happens during model design.*

1.7   Comprehension and hands-on

This textbook aims to transmit a practical insight into deep learning methods, their potential, and their challenges. In our own research, we actively contribute to particle physics, astroparticle physics, solid-state physics, and therefore will present beyond simple educational applications advanced examples primarily from these fields.

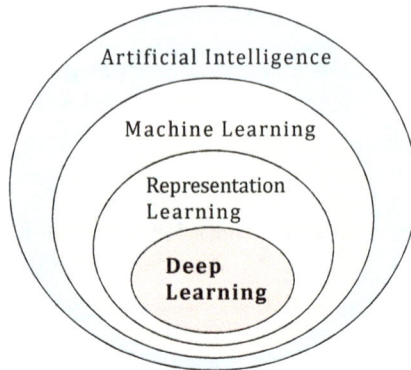

Fig. 1.4. Machine learning is a subcategory of artificial intelligence (AI), as is representation learning and deep learning.

For understanding this textbook, we expect our readers to have reached an advanced bachelor level or the masters level in physics education and have basic knowledge in linear algebra, statistics, functions, and mappings. We also expect our readers to code simple programs on their computers and make themselves familiar with the *Python* language, if not already done [6]. Readers who require theoretical knowledge about deep learning methods are referred to computer science books such as Ref. [1].

The order of the book corresponds to a possible course program that can be adjusted according to the prior knowledge of the participants:

(1) In this part, *Deep Learning Basics*, we will explain the elementary components of neural networks and their interplay. We will show how the network parameters can be optimized and how the optimization quality can be controlled.

(2) In the part on *Standard Architectures of Deep Networks*, we show variants of networks tailored to the input data's specific properties. Such properties are, for example, translation, rotation, or permutation symmetries. The consideration of such properties in the architecture frequently leads to a performance increase of the networks.

(3) In the part *Introspection, Uncertainties, Objectives*, we present methods to gain insight into how networks work and show which input they are most sensitive to. Furthermore, we explain how to quantify the uncertainties that arise when analyzing data with

deep networks. Finally, we will discuss variants of objective functions and criteria for their selection.

(4) In the part *Deep Learning Advanced Concepts*, we enter new territory by exploring the potential of data-driven knowledge discovery that is emerging with deep learning. We discuss different supervision levels during the optimization of network parameters. We also explain configurations with multiple networks working with or against each other.

All examples and exercises presented in this book are based on the Python language [6] and use the Keras software [7–9], which itself is an easy-to-use interface to the program library TensorFlow [10] (Fig. 1.5). We build networks, optimize their parameters and evaluate their performance. Note that there are several alternative open-source libraries, e.g. PyTorch [11], for building neural networks and we have selected Keras as a particularly straightforward gateway.

```python
from tensorflow import keras
layers = keras.layers
# setup and train a simple regression network with Keras
model = keras.models.Sequential()
model.add(layers.Dense(4, activation='relu', input_dim=2))
model.add(layers.Dense(1, activation='sigmoid'))
model.compile(loss='MSE', optimizer='SGD', metrics=['accuracy'])
model.fit(xdata, ydata, epochs=200)
```

Fig. 1.5. All examples and exercises in this book are written in the programming language Python using the Keras library to build neural networks.

Summary 1.1.

(1) Scientific questions to be answered from data require corresponding algorithms. Machine learning denotes a procedure to obtain suitable algorithms for data analyses. A 'machine' consists of multiple mappings that can be individually adjusted by parameters. To optimize these parameters, a data set is needed, and an objective function that is to be minimized.

(2) Overall, physicists' core tasks are to select, improve or develop network architectures suited to answer a scientific question, formulate objective functions that succeed in adapting millions of network parameters, understand the network's functionality, and quantitatively measure the statistical and systematic uncertainties of network outputs.

Chapter 2

Models for data analysis

Scope 2.1. In this chapter, we first recall different types of data that we use in physics analyses. Then, we discuss models for answering scientific questions from the measurement data. Classically, data analysis is based on physics laws from which a model for the interpretation of the data is formulated.

The way for data-driven model building works quite differently. Here, neural networks themselves are models that can be flexibly configured by adjusting their parameters. How the neural network model is designed depends, on the one hand, on the data used to adapt the parameters. On the other hand, an objective function is employed to compare the network's output with an expected value. Using a simple interpolation example, we show how a network model is designed.

2.1 Data in physics research

Several data types to be used for input into neural networks are presented in Fig. 2.1. Later in the textbook, we will discuss various possibilities for presenting these data as input to a network. Here, it is sufficient to imagine a vector formed from all relevant input values. The vector is passed as input to the network (Fig. 1.3).

To exemplify different data types for data analysis, we describe here functionalities of modern mobile phones. With their many sensors, they have become portable mini-laboratories. After that, we extend to examples taken from realistic physics experiments.

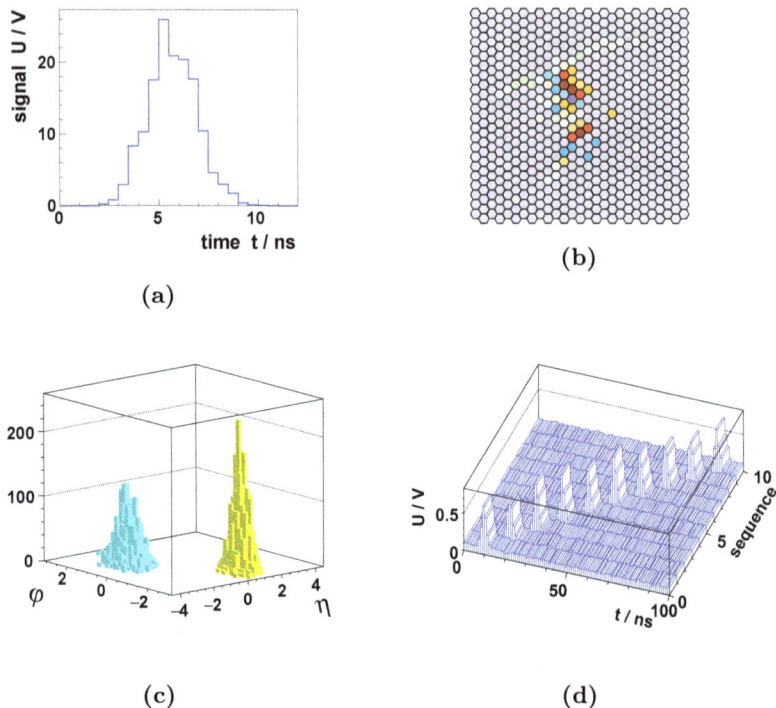

(a)

(b)

(c)   (d)

Fig. 2.1. Examples of data types: (a) sensor amplitude as a function of time, (b) camera image of a telescope, (c) particles of a collision projected onto a physics-motivated coordinate system, (d) sequence of short pulses.

2.1.1 *Data from events*

A general situation consists in a data set of k events, where always the same n measurement variables are recorded. For each event, we can consider the measured variables as an n-dimensional vector $\vec{x}$. Together, variables and events comprise a table with n columns and k rows.

2.1.2 *1D-sequences (time series)*

A one-dimensional sequence of a single variable x yields an *ordered list of numerical values* [Fig. 2.1(a)]. This can be extended to an n-dimensional vector $\vec{x}$ holding n different variables.

Examples from mobile phones are time-ordered series of electric amplitudes related to the microphone and the loudspeaker for recording and playing audio. Many mobile phones also offer additional sensors such as a pressure sensor, a light sensor, a temperature sensor, and three-dimensional acceleration sensors. With the GPS receiver for positioning, two-dimensional location and speed determination are possible. All sensors provide numerical values as a function of time and can be read out comfortably, e.g. with the *phyphox*[1] app to be used for physics experiments at school and university.

2.1.3 *Regular grid (image type data)*

Data on a regular grid are two-dimensional or even higher-dimensional arrangements of numerical values.

With its camera, a mobile phone provides a semiconductor with several million pixels arranged on a two-dimensional Cartesian grid to provide intensity information.

Many professional instruments used in physics are, of course, on a much higher qualitative and quantitative level than mobile phones. One example is telescopic cameras in the research field of astronomy, whose sensors are often arranged on hexagonal structures instead of Cartesian grids and which have a much higher quantum efficiency for detecting even the smallest light intensities [Fig. 2.1(b)]. But the principle of generating data of different types and delivering numerical values is similar.

In particle physics experiments located at an accelerator, e.g. the multi-purpose detectors ATLAS and CMS at CERN, the beam particles collide causing creation of many particles. A special projection of such data onto a regular grid is shown in Fig. 2.1(c).

2.1.4 *Regular grid combined with 1D-sequences*

The combination of the above-discussed cases yields *sequences* of data presented on *regular grids* [Fig. 2.1(d)]. A prime example is videos containing time-ordered series of images.

2.1.5 *Point cloud (3D, spacetime)*

A three-dimensional form or object can be described by a set of data points in space. Each point is assigned a vector to locate its position. All vectors form a so-called *point cloud*. For example, temperature measurements at

---

[1] https://phyphox.org

different locations in a three-dimensional space can naturally be described in terms of a point cloud.

2.1.6   *Heterogeneous data (ensembles of devices)*

Often not only individual devices are used, but entire *ensembles of devices*, each of which provides essential data and thus contributes to an overall measurement. One example is the distributed sensors for seismic measurements. The temporal and spatial relationships of signals from many such devices provide information about an earthquake.

Another example of the combination of many devices is the beam monitors of a particle accelerator. The exact positions of the particle bunches in the accelerator can be followed along the accelerator.

The detectors ATLAS and CMS at the CERN accelerator consist of millions of sensors to detect newly created particles as they pass through. The sensors comprise numerous different technologies and all together form a highly heterogeneous system.

2.2   Model building

The scientific question under investigation determines which information from possibly many sensors should be selected. Based on our intuitive understanding of physics, we often have expectations about the aspects of the measured data that are relevant for solving a task.

Example 2.1. Physicist's model for locating the earthquake epicenter: An earthquake typically has a center from which waves propagate in all directions and reach the stations for seismic measurements sequentially (Fig. 2.2). This experience is built into the *physics model* which we use to interpret the measurement data and answer the questions, e.g. where the earthquake's epicenter was located.

The velocities of compression waves v_p (longitudinal to the propagation direction of the wave) and shear waves v_s (transverse to the propagation direction) are known. Using the signal times t_i of several measuring stations i and the station distances x_i, linear models $t = x/v_s$ or $t = x/v_p$, or a combination of both, can be used to determine the location of the epicenters (Fig. 2.3).

The procedure is different for data-driven approaches. *Here the neural*

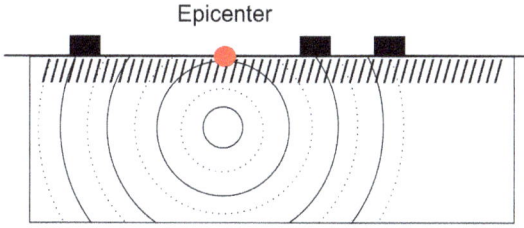

Fig. 2.2. Origin and epicenter of an earthquake, the circles represent compression waves and shear waves which propagate with different velocities to the measurement stations.

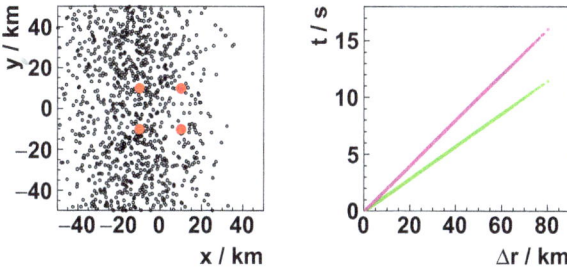

Fig. 2.3. Earthquake: (left) locations of epicenters (black symbols) and four measurement stations (red symbols), (right) arrival times of compression waves and shear waves depending on the distance to the epicenter.

network itself becomes the model (Fig. 1.3). We do not need to specify a physically motivated relation such as the linear relation $t = x/v$. Rather, the network is supposed to learn it from training data.[2] In principle, a linear model can be formed in the network by adjusting the network parameters accordingly. But the real strength of networks is that they can be adjusted to yield extremely complex, nonlinear models by using training data.

Thus, to make the neural network a useful model, we need a data set to adjust the network parameters. To enable evaluation of the model quality, we will need an additional data set that is statistically independent of the first data set. We will see in Sec. 5.2 that we need to divide the data set into three parts and use the first and second partial data sets for adjusting the network model, the third data set for checking the result.

---

[2]However, in practice, using known physical laws in the construction of the model can be helpful.

We require the data set to contain an expected value for adjusting the network parameters that one can compare with the network prediction. The comparison is realized with an objective function that provides a quantitative distance measure between the network prediction and the expected value. The network parameters are optimized by minimizing the distance measure of the objective function.

If we then take a statistically independent data set, we can check whether the network model works correctly on data that the network has never processed before. This is referred to as the *generalization capability* of the network.

Example 2.2. Data-driven model for locating the earthquake epicenter: In the above example of earthquake events, the data set should contain the measurement data of the seismic stations and the position of the epicenter (latitude b, longitude l). To the network, we give only the measurement data of the stations as input information and make sure that the order of the stations i is always the same when entering the arrival times of the compression waves $(t_{i,p})$ and shear waves $(t_{i,s})$ into the network. It is then not necessary to enter the position of the stations in the network, nor is it necessary to enter information on typical propagation velocities of compression and shear waves.

We require two output values of the network, the latitude $\hat{b}$ and the longitude $\hat{l}$ of the epicenter event-wise. We compare this output $(\hat{b}, \hat{l})$ of the network with the true value (b, l) of the data set. In an iterative learning process of the neural network with the data, we demand that the network's model with its output values adapts better and better to the epicenters' correct positions. In this way, the neural network becomes the model to predict the epicenter from the data of the seismic stations (see Sec. 7.2).

2.3   Data-driven model optimization

Neural networks are supposed to learn from data to perform a specific task. The *learning* process technically involves the adjustment of thousands or even millions of network parameters that need to be set appropriately. The *optimization* of the *network parameters* is also referred to as *network training*. We aim here for a single output value f to be predicted by the network based on a set of input values $\vec{x}$.

For a straightforward network training, the data set should contain the input values $\vec{x}$ and, in addition, a target value y, which the network ideally should predict. For many physics problems, such data sets can be obtained by simulations where the target values y are usually known. Alternatively, one can use control measurements or other algorithms to obtain data with known target values. This target value y is also referred to as *label*. Data carrying a label are called *labeled data*.

Depending on their scientific question, physicists choose an *objective function* $\mathcal{L}$ for quantifying a distance measure between the network prediction $f(\vec{x})$ and the target value y. Note that the target value y is only used in the objective function but never given to the network as input.

A common way forward for model optimization is to take k residual values $\Delta = |f(\vec{x}) - y|$ between the network prediction and the target value. The objective function $\mathcal{L}$ averages the squared residuals $\mathcal{L} = (1/k) \sum_{i=1}^{k} \Delta^2$. The optimization of the network parameters is then performed by minimizing the objective function $\mathcal{L}$.

Example 2.3. Model building for interpolating data: An example of the optimization process is shown in Fig. 2.4. The trajectory of the points (data pairs (x, y)) is to be interpolated through a network consisting of three hidden layers with 30 nodes each. The output of the network is denoted as $f(x)$. Before training, the network parameters were set randomly, so the initial output had nothing to do with the interpolation of the points.

The objective function is the mean value of the squared residuals, i.e. the squared differences between the value y_i of the data pair and the output value $f(x_i)$ of the network. We optimize the network parameters by minimizing this objective function $\mathcal{L}$.

This process can be observed in Fig. 2.4. For the optimization of the network, a few hundred iterations are needed during which the training data is used several times. The term *epoch* describes a training cycle where all available data are used once for setting the weights. After the first cycles, the output $f(x)$ of the network is not yet satisfactory. Still, already after 800 epochs, $f(x)$ corresponds quite well to an interpolation of the points: The network parameters are now well adjusted. The improvement is also visible in the evolution of the objective function in the lower figure.

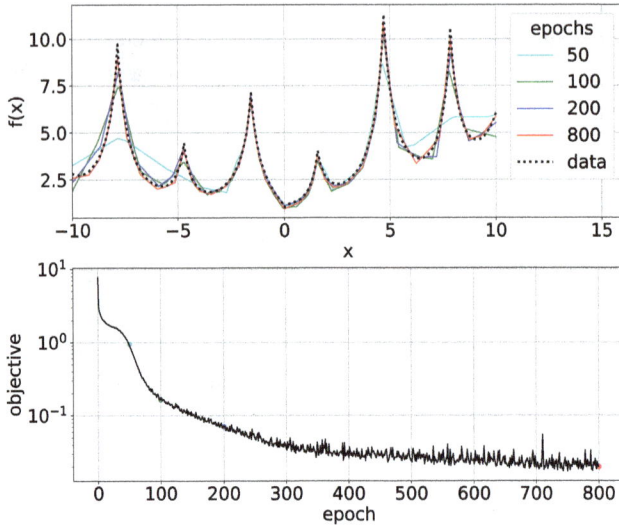

Fig. 2.4. (Top) Optimization (training) of a network with the goal to interpolate the black points (x, y) by a function. The function is implemented here by a network with one input value x and one output value $f(x)$. (Bottom) When optimizing the network parameters, the differences of the $f(x)$ output values and the target values y of the point pairs are minimized using an objective function. With more training cycles the network parameters improve and thus the interpolation of the point pairs through the network.

This example with a single input variable x and an output $f(x)$ is kept very simple. For typical problems, the networks have n input variables and have m output values. However, the principle of network training with an objective function remains the same and is followed even for more complex networks.

Summary 2.1.

(1) A variety of different data types are handled in physics data analyses. Typical examples are:

   (a) A data set, simply consisting of k events each with n measurement variable,

   (b) One-dimensional sequences of numerical values. A typical example is a time series, which contains time-ordered amplitude measurements of a sensor,

   (c) Regular grids containing numerical values. An example is an image containing intensity measurements arranged on a regular grid,

   (d) Combinations of regular grids with 1D-sequences. Examples are videos, i.e. images arranged in a time sequence,

   (e) Point cloud describing a three-dimensional form or object. Each point is assigned a vector to locate its position.

   As for now, imagine these data as values of a vector for input to the network. Various alternative ways of feeding such data into the network will be discussed later in detail. Also, note that the data set may consist of a collection of equal or variable length events.

(2) Models pose a scientific question to the data. Neural networks are models that can be flexibly adjusted by their parameters. Essential for the specification of a model is the information contained in the data sets for training the network.

(3) The objective function determines which aspects of the model are particularly emphasized in the data. It fulfills a central guiding task for the optimization of the network parameters.

Chapter 3

Building blocks of neural networks

Scope 3.1. This chapter explains the basic components of a neural network used to create a model for data analysis. At each node of a network, there is a linear transformation plus displacement (i.e., an affine mapping) and a nonlinear transformation by a so-called activation function. To specify the neural network model, the linear transformation and displacement parameters are freely adjustable. Activation functions ensure that the model has nonlinear properties. The network output can be one or more values. The network's task can be to predict continuous values in response to input values (regression), or to predict unambiguous assignment of the input data into different categories (classification). Furthermore, we explain the Universal approximation theorem.

3.1 Linear mapping and displacement

In Fig. 3.1, we show a network with two input values $\vec{x} = (x_1\, x_2)^T$. These inputs are combined via two *nodes* in the *hidden layer* to one output value z. The mathematical operation of the first hidden layer is a linear transformation plus displacement (so-called *affine transformation*) with a matrix $\mathbf{W}$ of *weight* parameters W_{ij} and a vector $\vec{b}$ of displacement parameters b_i named *bias*:

$$\vec{y} = \mathbf{W}\,\vec{x} + \vec{b} \tag{3.1}$$

$$\begin{pmatrix} y_1 \\ y_2 \end{pmatrix} = \begin{pmatrix} W_{11} & W_{12} \\ W_{21} & W_{22} \end{pmatrix} \begin{pmatrix} x_1 \\ x_2 \end{pmatrix} + \begin{pmatrix} b_1 \\ b_2 \end{pmatrix}$$

$$= \begin{pmatrix} W_{11}\,x_1 + W_{12}\,x_2 + b_1 \\ W_{21}\,x_1 + W_{22}\,x_2 + b_2 \end{pmatrix} \tag{3.2}$$

$$y_1 = W_{11} x_1 + W_{12} x_2 + b_1$$

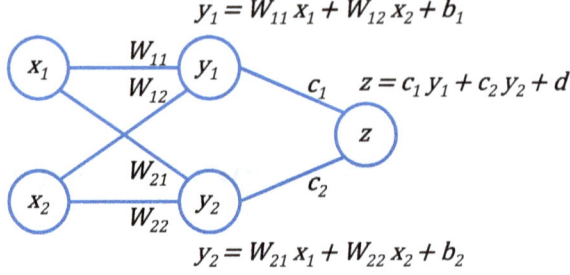

$$z = c_1 y_1 + c_2 y_2 + d$$

$$y_2 = W_{21} x_1 + W_{22} x_2 + b_2$$

Fig. 3.1. A network of linear mappings plus displacements.

Using the results from the hidden layer (3.2), we perform another affine transformation using the weights $\mathbf{W'} = (c_1\ c_2)$ and the bias d to obtain the output value z of the network (Fig. 3.1):

$$z = \mathbf{W'}\,\vec{y} + d$$

$$= (c_1\ c_2) \begin{pmatrix} y_1 \\ y_2 \end{pmatrix} + d$$

$$= (c_1\ c_2) \begin{pmatrix} W_{11} x_1 + W_{12} x_2 + b_1 \\ W_{21} x_1 + W_{22} x_2 + b_2 \end{pmatrix} + d$$

$$= c_1 W_{11} x_1 + c_1 W_{12} x_2 + c_1 b_1 + c_2 W_{21} x_1 + c_2 W_{22} x_2 + c_2 b_2 + d$$

$$= \underbrace{(c_1 W_{11} + c_2 W_{21})}_{\equiv W_1''} x_1 + \underbrace{(c_1 W_{12} + c_2 W_{22})}_{\equiv W_2''} x_2 + \underbrace{c_1 b_1 + c_2 b_2 + d}_{\equiv b''}$$

$$= W_1'' x_1 + W_2'' x_2 + b'' \tag{3.3}$$

The result of a series of affine mappings is again an affine mapping. This also applies to networks with additional hidden layers. Models formed with such a network could only describe linear phenomena and are therefore not suitable as general modeling tools.

3.2   Nonlinear mapping: activation function

To describe nonlinear problems with networks, a nonlinear function σ is applied at each node after the affine mapping (3.1) as shown in Fig. 3.2:

$$\sigma(\vec{y}) = \begin{pmatrix} \sigma(y_1) \\ \sigma(y_2) \end{pmatrix} \tag{3.4}$$

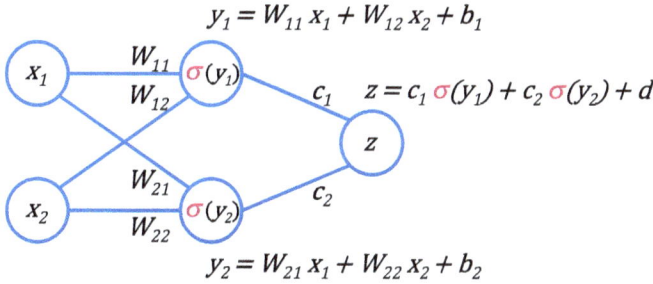

Fig. 3.2. The forward pass in a neural network consists at each node of two operations: (1) an affine mapping (linear mapping plus displacement), (2) the application of an activation function σ.

In analogy to neurons, σ is called *'activation'* function. When calculating the output value, a mathematical simplification as in Eq. (3.3) is then no longer possible:

$$z = \mathbf{W}' \, \sigma(\vec{y}) + d$$

$$= (c_1 \; c_2) \begin{pmatrix} \sigma(y_1) \\ \sigma(y_2) \end{pmatrix} + d$$

$$= c_1 \, \sigma(W_{11} \, x_1 + W_{12} \, x_2 + b_1) + c_2 \, \sigma(W_{21} \, x_1 + W_{22} \, x_2 + b_2) + d \quad (3.5)$$

An example of an activation function is the so-called rectified linear unit (ReLU) function [Fig. 3.3(a)], which passes positive input values unchanged to the next network level but sets negative input values to zero. When using this function, the displacement parameters b_1 and b_2 have an obvious meaning: they define threshold values, above which the result is passed on [Fig. 3.2 and (3.5)]. The ReLU is one of the most commonly used activation functions as it is computationally inexpensive and delivers fine results.

Several other activation functions can be chosen. For example, the sigmoid [Fig. 3.3(b)] was one of the first introduced activation functions due to its similarity to neuroscience. The hyperbolic tangent, shown in Fig. 3.3(c), is a further example.

3.3 Network prediction

At each node the two above described mappings are performed. First, the affine transformation $\vec{y} = \mathbf{W} \, \vec{x} + \vec{b}$ is calculated with the freely adjustable parameters $\mathbf{W}$ and $\vec{b}$ (3.1). On the result $\vec{y}$ the nonlinear activation function

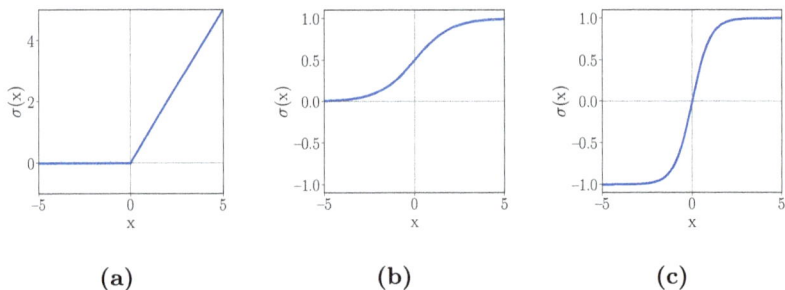

(a)   (b)   (c)

Fig. 3.3.   Examples of activation functions, (a) ReLU, (b) sigmoid, (c) tanh.

$\sigma(\vec{y})$ is applied (3.4). For each node the combined operation is given by:

$$\vec{z} = \sigma\left(\mathbf{W}\,\vec{x} + \vec{b}\right) \tag{3.6}$$

As seen in Fig. 1.3, neural networks can have many layers with many nodes. In each layer, the information from the previous layer is taken as input variables. The prescription of calculating the network output from the input values respecting all intermediate nodes of the hidden layers represents the *forward pass* in a neural network.

The output of a network is also referred to as *network prediction*. Depending on the physicist's question, different tasks can be solved.

3.3.1 *Regression*

In Fig. 2.4, we interpolated data points using a network. The network receives on input the value $x \in \mathbb{R}$, a real number, and delivers the output value $z \in \mathbb{R}$, also a real number. The network, therefore, works like a function $z = f(x)$, which returns a continuous value in response to a real input value. A task that yields such continuous predictions is called *regression* [see Fig. 3.4(a)].

The adaptation of the network parameters is called *network training*. Here, it works like a fit to the data.

The example shown in Fig. 3.4(a) is kept very simple because the network has only $n = 1$ input variable and $m = 1$ output value. The extension to networks with $n = 2$ input variables can be imagined using Fig. 3.2, where the two inputs are combined to one output value.

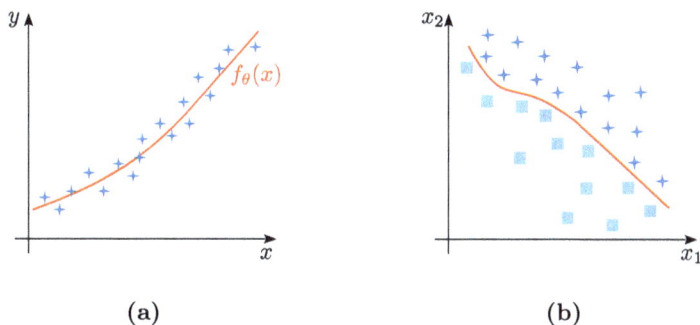

Fig. 3.4. Model predictions for (a) regression and (b) classification tasks.

Example 3.1. Prediction of car velocity: Given measurements $\vec{x}$ of a car's distance traveled $x_1 = s$ (for example, the odometer reading) at time $x_2 = t$, we can use a network to predict its driving velocity $z = v$.

Example 3.2. Prediction of earthquake properties: For the prediction of the arrival time t of an earthquake at location (l, b) we can use as input values the radio-transmitted information of n measuring stations and let the network calculate the prediction of the arrival time $z = t$ at this location.

In a variation of the network with $m = 2$ output values, we ask for the epicenter's coordinates from the information of n measurement stations. Here the network calculates a prediction for the latitude $z_1 = l$ and the longitude $z_2 = b$ of the epicenter (also refer to Secs. 2.2 and 7.2).

3.3.2 *Classification*

Another common variant of network output is the unambiguous assignment of input data into different categories called classification. For example, if we wish to separate signal data from background data, we have two categories, shown in Fig. 3.4(b) as cross symbols and box symbols. Separation with the x_1 variable alone is impossible, nor does it work with the x_2 variable. What is required is a separation line from a combination of x_1 and x_2 as shown in the figure.

We again use a network like in Fig. 3.2 with $n = 2$ input values x_1, x_2

and the output value z, i.e. output dimension $m = 1$. Now comes a common trick. The output z is additionally subjected to a sigmoid function $z' = \sigma(z)$ [Fig. 3.3(b)] with the new output value $z' \in [0, 1]$. The network training is then carried out so that signal data is led to output values $z' > 0.5$ and background data to output $z' < 0.5$. This builds the desired dividing line at $z' = 0.5$ in the network. Finally, the network can be used to evaluate data that has not been used to train the network.

As for regression, a classification network can be extended to many input variables $n \gg 1$. The network can also be extended to many output categories ($m > 2$) often called *classes*, with the network assigning each input object to a category. A frequently used example is the classification of natural images into categories such as animals, vehicles, buildings, etc.

3.4 Universal approximation theorem

The general question is, can a neural network represent any function that is needed for the modeling of physics-related problems? Are the multiple nodes with their affine transformation and activation functions sufficiently flexible for this?

The Universal approximation theorem provides the positive answer [12, 13]. The theorem was developed for the approximation of continuous functions on closed and bounded subsets of the Euclidean space $\mathbb{R}^n$, which fortunately comprises typical functions relevant to physics:

> *A feed-forward network with linear output and at least one hidden layer with a finite number of nodes can approximate any of the above functions to arbitrary precision.*

Meanwhile, the Universal approximation theorem has been confirmed for multi-layer networks with a limited number of nodes and various activation functions, e.g. the ReLU function [14–16].

However, the learnability of the function remains an open question. The Universal approximation theorem says nothing about a suitable learning procedure that can be used to adjust the network's parameters accordingly. From experience, one can state that multi-layer networks with a limited number of nodes can be trained rather well.

3.5 Exercises

Exercise 3.1. Navigate to `www.deeplearningphysics.org`. This webpage hosts additional material for most exercises, as well as example solutions. Whenever we refer in the following to the *exercise page*, this is the site to be visited.

Exercise 3.2. Before diving into deep neural networks, let us first use simple linear methods to solve a regression task. We will predict the quality of Portuguese white wine — so-called *vinho verde* — from its physicochemical properties. The quality of a wine is, of course, a somewhat subjective quantity. In this dataset, the quality assessment by human tasters and more easily measurable chemical quantities such as acidity are provided. We will explore how well this quality assessment can be explained from objective measurements alone.

(1) Use the Notebook (Exercise 3.2) from the exercise page to download and import the data.

(2) First, we want to understand the dataset better. Plot the distribution of each of the features for the training data as well as the 2D distribution of each feature versus quality. Also, calculate the correlation coefficient for each feature with quality. Which feature seems most predictive of the quality?

(3) Calculate the linear regression weights by finding the weight matrix $\mathbf{W}$ that minimizes the mean squared error (MSE):

$$(\mathbf{W}\,\vec{x} - \vec{y})^2. \tag{3.7}$$

Here $\vec{x}$ are the training data chemical properties and $\vec{y}$ the true quality labels.

(4) Use the weights to predict the quality for the test dataset. How does your predicted quality compare with the true quality of the test data? Calculate the correlation coefficient between predicted and true quality and draw a scatter plot.

Exercise 3.3. Checkerboard: Familiarize yourself with the basic building blocks of a neural network.

Open the Tensorflow Playground (`playground.tensorflow.org`) and select on the left the checkerboard pattern as the *data* basis. The data is taken from a two-dimensional probability distribution and is represented by the value pairs x_1 and x_2. The regions $x_1, x_2 > 0$ and $x_1, x_2 < 0$ are shown by one color. For value pairs with $x_1 > 0, x_2 < 0$ and $x_1 < 0, x_2 > 0$, the regions are indicated by a different color.

In *features*, select the two independent variables x_1 and x_2 and start the network training. The network learns that x_1 and x_2 are for these data not independent variables, but are taken from the probability distribution of the checkerboard pattern.

(1) Try various settings for the number of layers and neurons using ReLU as activation function. What is the smallest network that gives a good fit result?
(2) What do you observe when training networks with the same settings multiple times? Explain your observations.
(3) Try additional input features: Which one is most helpful?

Summary 3.1.

(1) At each node of a neural network two transformations are performed. First, all input values are combined by a linear transformation and displacement (i.e. an *affine transformation*): $\vec{y} = \mathbf{W}\,\vec{x} + \vec{b}$. The weight parameters $\mathbf{W}$ and the bias parameters $\vec{b}$ can be adapted for solving the desired challenge. In order to model nonlinear phenomena with neural networks, a nonlinear function, the so-called *activation function* $\sigma(\vec{y})$, is applied to the result $\vec{y}$. The parameters $\vec{b}$ adjust the switching properties of the activation function.

(2) The output of a network is also referred to as *network prediction*. Depending on the problem, the network can be used as a high-dimensional function approximator, which provides for n input values m continuous output values. This procedure is referred to as *regression*.

When using a network to classify objects into m categories, the network output is normalized by a suitable function to enable the object's unique assignment to a discrete category. The assignment of objects to categories is called *classification*.

(3) The *Universal approximation theorem* states that a feed-forward network with linear output and at least one hidden layer with a finite number of nodes can, in theory, approximate functions of interest in physics to arbitrary precision. Extensions of the theorem also show validity for multi-layer networks with a limited number of nodes. Still, there is no guarantee: learning the desired function through network training remains a challenge in each scientific application.

Chapter 4

Optimization of network parameters

Scope 4.1. In this chapter, we learn what is required to train a network. We show how to prepare the input data for networks and explain why it requires care. We give methods for initializing the network parameters. For regression and classification tasks, we provide typical objective functions used in the optimization process. We detail backpropagation and the stochastic gradient descent method to optimize the parameters. Finally, we introduce dynamically adjusted step sizes for the optimization process.

4.1 Preprocessing of input data

Numerical stability is crucial for the successful building of a neural network model. If the ranges of the input data $\vec{x}$ vary by many orders of magnitude and are different for the individual components x_i, adjusting network parameters becomes difficult. Careful preparation and pre-processing of the input data are essential for successful model building. The following criteria should be checked.

Zero-centered: The gradient of the widely used ReLU activation function $\sigma(\vec{y})$ (3.4) of a network node changes significantly close to zero (Fig. 3.3), where it increases or reduces the importance of the results $\vec{y}$ of the affine mappings ($\vec{y} = \mathbf{W}\vec{x} + \vec{b}$ (3.1)). Therefore, input variables x_i are often subtracted by their mean value $\langle x_i \rangle$ so that they cross over positive and negative values:

$$x_i - \langle x_i \rangle \tag{4.1}$$

Order of magnitude: If the numerical values of some of the input variables x_i are small, but other input variables x_k have huge values, the large values could be preferred in the network training. Here, the variables should first be brought to similar orders of magnitude. If input data vary over a wide range, mean value $\langle x_i \rangle$ and standard deviation σ_i of each variable are used:

$$x_i' = \frac{x_i - \langle x_i \rangle}{\sigma_i} \tag{4.2}$$

For values in a fixed interval, minimum $x_{i,\min}$ and maximum $x_{i,\max}$ can be helpful scaling factors:

$$x_i' = 2 \, \frac{x_i - x_{i,\min}}{x_{i,\max} - x_{i,\min}} - 1 \tag{4.3}$$

Logarithm: If the input variable x_i fluctuates by many orders of magnitude and the transition range between small and large values is important for the problem to be solved, it can be useful to distribute the data more evenly by a monotonous function. Typically used functions are the logarithm or the exponential function:

$$x_i' = \log x_i \tag{4.4}$$
$$x_i' = e^{-x_i} \tag{4.5}$$

Decorrelation: If input variables x_i and x_j are strongly correlated, it can be helpful for efficient network training to first perform a variable transformation to decorrelate the input variables.

Global normalization: For series of variables containing the same measurement quantity, e.g. pixel intensities of an image, normalizations are performed globally using the methods described before. For example, the pixel intensities can be normalized with respect to their global average (4.1).

Example 4.1. Data normalization for seismic stations: In a thought experiment, we have collected a total of 100,000 earthquake events between 2015 and 2020 with 50 measuring stations that record the seismic movements as a function of time. The data of each station i are already pre-processed in such a way that they only contain the time $t_{p,i}$ of the maximum amplitude $A_{p,i}$ of the compression wave, and the time $t_{s,i}$ of the maximum amplitude $A_{s,i}$ of the shear wave (compare with Sec. 2.2).

4.1.1 *Normalization*

The times $t_{p,i}, t_{s,i}$ are measured in seconds following the Unix timestamp, which was started on January 1, 1970 at 00:00 o'clock of the Universal Time Coordinated (UTC) and therefore exceeds one billion seconds. Operating with large numerical values challenges the precision of the numerical calculations inside the network, which can be avoided. A practical approach is to normalize the times $t_{p,i}$ in each event using the first and last time recordings (4.3):

$$t'_{p,i} = 2\,\frac{t_{p,i} - t_{p,i,\min}}{t_{p,i,\max} - t_{p,i,\min}} - 1 \qquad (4.6)$$

This results in values between $-1 \leq t'_{p,i} \leq 1$. Repeat the same transformation for the times $t_{s,i}$.

The maximum amplitudes are assumed to be distributed between numerical values $[0, 1000]$, with the vast majority of the amplitudes being close to zero. If the transition range between small and large amplitudes is of interest for the physics problem to be solved, it may be useful to distribute the data more evenly using the logarithm (4.4):

$$A'_{p,i} = \log A_{p,i} \qquad (4.7)$$

The same transformation is applied to the amplitudes $A_{s,i}$.

4.2 Epoch and batch

A network training is performed iteratively. The available training data is used repeatedly and as efficiently as possible. Two terms are crucial to describe this procedure: epoch and batch.

Epoch: One epoch of a network training denotes the one-time, complete use of all training data. A network training usually consists of many epochs, i.e. multiple reuses of all training data.

Batch, minibatch: The parameters of a network are iteratively optimized in many small steps. Using all training data in each step would be time-consuming and highly inefficient. Using instead a randomly selected, sufficiently large subset of the training data in each iterative step has proven to be a good practice. The subset of data is referred to as a *'minibatch'*, or a bit sloppily as *'batch'*. Often, *batch sizes* are chosen in powers of 2^m, e.g. $k = 32$.

The exact size of the batch to be chosen depends on the problem to be solved and impacts the quality of the model building. For deciding on the size k, computing costs proportional to k need to be balanced with precision proportional to $1/\sqrt{k}$ for averaging over k values. Extreme choices may lead to disadvantages: large or too-small batch sizes, e.g. $k = 8{,}192$ or $k = 1$, may lead to inefficient and time-consuming training or compromise the generalization performance of the network (see Sec. 5.1).

Progress in the network training is typically evaluated in terms of the number of epochs. The total number of iteration steps due to batches is much larger, which can be seen by dividing the number d of training samples by the batch size k.

4.3 Parameter initialization

When choosing the network parameters' initial values, the primary goal is breaking the symmetry of the weight parameters. This is to avoid different network nodes learning identical mappings instead of the desired diverse mappings $\vec{y} = \mathbf{W}\,\vec{x} + \vec{b}$ (3.1) within the network. Symmetry breaking of the initial weights $\mathbf{W}$ is achieved by random numbers taken from a uniform $[-s, s]$ or a Gaussian distribution with mean value $\mu = 0$ and standard deviation σ. The bias parameters $\vec{b}$ are usually set to zero. Besides, the initial weights $\mathbf{W}$ should neither be too large nor too small. A layer-wise random initialization with a standard deviation scaled with the number of input nodes n_{in} and the number of output nodes n_{out} has proven favorable for the optimization process described below. As the variance σ^2 for Gaussian-distributed initialization of layers with activation function tanh [Fig. 3.3(c)], it is recommended to use the *Glorot normal* or *Xavier normal* initialization [17]:

$$\sigma^2 = \frac{2}{n_{\text{in}} + n_{\text{out}}} \tag{4.8}$$

Taking the simplified view of equal numbers $n_{\text{in}} = n_{\text{out}}$, it is firstly achieved that for input values of the order of 1 the variance V of the summed output values of a network layer, which serve as input for each node in the subsequent layer, is $V = \sum_{i}^{n_{\text{in}}} \sigma_i^2 \approx 1$. Second, the variance of the gradient sum is also $V = \sum_{j}^{n_{\text{out}}} \sigma_j^2 \approx 1$, which is calculated from the perspective of the n_{out} output nodes of the network layer (see backpropagation in the following Sec. 4.5).

When using the ReLU activation function (Fig. 5.8), clipping input values below zero causes half of all output values to be lost. Here, a Gaussian-distributed initialization is recommended with the variance [18]:

$$\sigma^2 = \frac{2}{n_{\text{in}}} \tag{4.9}$$

For initialization of the weight parameters the uniform distribution can be used as well. For symmetric intervals around zero, its standard deviation denotes $\sigma = (s - (-s))/\sqrt{12} = |s|/\sqrt{3}$. For obtaining the same spread of the weights $\mathbf{W}$ as for the Gaussian initialization, the corresponding range is $s = 3\sigma$ using σ from (4.8) or (4.9), respectively.

4.4 Objective function

The *objective function* is the heart of the optimization process and thus of the network modeling. This function represents a global measure evaluating the quality of the network prediction. Depending on the user community, the objective function is also called 'cost' or 'loss'.

When training a network, thousands or even millions of parameters must be set correctly. Successful optimization of the network parameters corresponds to reaching, or at least successfully approaching, a good local minimum on the hypersurface of all parameters. Like a fit with several parameters, the objective function gives each iteration a single scalar measure of how well the parameters have been adjusted so far. The minimum of the objective function is taken to indicate that a sufficiently good local minimum of all parameters has been reached. Whether this has actually been achieved must be verified in each individual project by evaluating the network predictions.

Below we present two commonly used objective functions, one for regression and another one for classification. For the network optimization, both require data with labels, i.e. data with an expected target value to be compared to the network prediction. This way of network optimization is referred to as *supervised training*. Later in Chapter 14, we will consolidate our understanding of objective functions and mention alternatives. In Part 4, we will also discuss network optimizations using data without labels.

4.4.1   *Regression: mean squared error (MSE)*

For tasks that require the network to predict one or more continuous variables, the *mean squared error* or short *MSE* is often used as the objective function $\mathcal{L}$ for optimizing the network parameters. To simplify matters, initially, we will consider an output value f of the network based on the input value x. MSE is the mean value (or expectation value $\mathbb{E}$) of the squared residuals between the network prediction $f(x)$ and the expected target value $y(x)$:

$$\text{MSE} = \mathbb{E}\left[(f - y)^2\right] \tag{4.10}$$

The objective function $\mathcal{L}$ given input values x, network predictions $f(x)$ and target values $y(x)$ all in one dimension [Fig. 4.1(a)] is explicitly calculated by:

$$\mathcal{L} = \text{MSE} = \frac{1}{k}\sum_{i=1}^{k}\left[f(x_i) - y(x_i)\right]^2 \tag{4.11}$$

For computing efficiency, estimates of $\mathcal{L}$ are obtained by using a minibatch, i.e. a sufficient number k of point pairs $(x_i, y(x_i))$ randomly selected from the total sample of available training data. Here, the index i denotes the position of the data in the minibatch, not a specific variable. For the optimum setting of the network parameters, the objective function should become minimal.

Extending the objective function to high dimensions n of the inputs $\vec{x}$ and dimension m of the network predictions $\vec{f}$ with components f_j and the target values $\vec{y}$ with components y_j is straightforward [Fig. 4.1(b)]:

$$\mathcal{L} = \frac{1}{k}\sum_{i=1}^{k}\sum_{j=1}^{m}\left[f_j(\vec{x}_i) - y_j(\vec{x}_i)\right]^2 \tag{4.12}$$

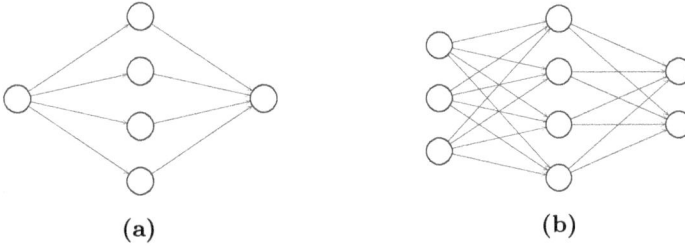

(a) (b)

Fig. 4.1. (a) Neural network with one-dimensional input and output, (b) neural network with three input nodes and two output nodes [19].

Example 4.2. Interpolation by neural network: In Fig. 2.4, we presented the interpolation of pairs of points (x_i, y_i) by a network and showed how, during the training, the network prediction improves with increasing epochs. During optimization, the mean squared residual values between the output values $f(x_i)$ of the network and the expected values y_i were calculated according to (4.11) using minibatches of point pairs (x_i, y_i).

4.4.2 *Classification: cross-entropy*

Neural networks are also used for classifying events or objects. To optimize a network for classification tasks, we need an objective function that processes probabilities.

Such probabilities remind us of statistical mechanics. The entropy definition according to Gibbs and Boltzmann is $S = k_B \log W$, where k_B is the Boltzmann constant and W stands for the total number of configurations in a system with a countable number of allowed configurations that are equally likely. We can relate this counting entropy to probabilities by recalling that the probability of each state is here $p = 1/W$, resulting in $S = -k_B \log p$.

If we generalize to the case where p is not constant, the term $\log p$ needs to be replaced by the expectation value $\mathbb{E}[\log p_j]$ over all probabilities p_j, i.e. average value $\langle \log p_j \rangle$, yielding

$$S = -k_B \langle \log p_j \rangle = -k_B \sum_j p_j \log p_j \tag{4.13}$$

which can also be used to measure entropy in a system *not* in equilibrium.

In information theory, entropy can be understood as a measure of ignorance. In this context, Shannon introduced $k_S = 1/\log 2$ instead of the

Boltzmann constant k_B as the normalization constant. This allows rewriting (4.13) in the form

$$S = -k_S \sum_j p_j \log p_j = -\sum_j p_j \log_2 p_j \qquad (4.14)$$

and thus measuring entropy in bits.

Here, we are interested in a related quantity, the cross-entropy. It distinguishes the true probabilities p_j from estimated probabilities q_j: $H = -\sum_j p_j \log q_j$. If the probability distributions of p_j and q_j coincide, H becomes minimal.

In a data set for training the network, the probabilities p are given by the true classification of the data objects[1] in m categories or classes. The estimated probabilities q are the predictions of the network. The objective function for network training is called *cross-entropy* and states:

$$\mathcal{L} = -\frac{1}{k} \sum_{i=1}^{k} \left[\sum_{j=1}^{m} p_j \log q_j \right]_i \qquad (4.15)$$

Note that i still gives the position of the data in the minibatch and j denotes the class or category. For efficient network training, estimates of $\mathcal{L}$ are computed using a minibatch of size k only. The objective function $\mathcal{L}$ should become minimal for optimized network parameters. For $m = 2$ alternative classes with probability sum $p_1 + p_2 = 1$ and a corresponding network prediction $q_1 + q_2 = 1$, cross-entropy simplifies to:

$$\mathcal{L} = -\frac{1}{k} \sum_{i=1}^{k} \left[p_1 \log q_1 + (1 - p_1) \log (1 - q_1) \right]_i \qquad (4.16)$$

To understand what is concretely transported in the variables p_j and q_j of (4.15), we will separate signal events from background noise in the following. The network works as before. We only adjust the network's output to contain the desired number of output nodes [Fig. 4.1(b)] and adapt the objective function to follow (4.15).[2]

As an activation function for the hidden layer we take here the tangent hyperbolic $\sigma(y) = \tanh(y)$ [Fig. 3.3(c)], which gives both positive and negative values. For the classification of events into two classes we use two nodes $\vec{z} \in \mathbb{R}^2$ as output of the network after the hidden layer:

$$\begin{pmatrix} z_1 \\ z_2 \end{pmatrix} = \begin{pmatrix} c_{11} & c_{12} \\ c_{21} & c_{22} \end{pmatrix} \begin{pmatrix} \sigma(y_1) \\ \sigma(y_2) \end{pmatrix} + \begin{pmatrix} d_1 \\ d_2 \end{pmatrix} \qquad (4.17)$$

---

[1] Usually discrete values 1 or 0.

[2] For only two classes, we can apply the trick with the sigmoid function presented in Sec. 3.3.2 requiring one output node only. Here we take the general perspective for working with $m \geq 2$ classes.

The results of this last layer are first amplified with the exponential function and normalized so that $\hat{z}_1 + \hat{z}_2 = 1$:

$$\vec{q} \equiv \begin{pmatrix} \hat{z}_1 \\ \hat{z}_2 \end{pmatrix} = \frac{1}{e^{z_1} + e^{z_2}} \begin{pmatrix} e^{z_1} \\ e^{z_2} \end{pmatrix}. \tag{4.18}$$

This is the so-called *softmax* function, which allows the results to be interpreted as probabilities $\vec{q}$. If we want to distinguish signal events from background noise, the network classifies inputs with $\hat{z}_1 > \hat{z}_2$ as signal and with $\hat{z}_1 < \hat{z}_2$ as noise.

The prerequisite for training this network is a data set containing distinct information for each event as to whether it belongs either to signal or background. We encode this information on the true value for each event in the vector:

$$\vec{p}^T = \begin{pmatrix} p_1 & p_2 \end{pmatrix} = \begin{cases} \begin{pmatrix} 1 & 0 \end{pmatrix} : \text{Signal} \\ \begin{pmatrix} 0 & 1 \end{pmatrix} : \text{Background} \end{cases} \tag{4.19}$$

This is often referred to as *one-hot-encoding*.

For training a network, we estimate the cross-entropy (4.15) for a minibatch of size k with data $\vec{x}_i$ consisting of $m = 2$ classes signal and background:

$$\mathcal{L} = -\frac{1}{k} \sum_{i=1}^{k} \left[\sum_{j=1}^{m=2} p_j(\vec{x}_i) \log q_j(\vec{x}_i) \right] \tag{4.20}$$

$$= -\frac{1}{k} \sum_{i=1}^{k} \begin{pmatrix} p_1(\vec{x}_i) & p_2(\vec{x}_i) \end{pmatrix} \begin{pmatrix} \log [\hat{z}_1(\vec{x}_i)] \\ \log [\hat{z}_2(\vec{x}_i)] \end{pmatrix} \tag{4.21}$$

$$= -\frac{1}{k} \sum_{i=1}^{k} \vec{p}^T(\vec{x}_i) \log \vec{q}(\vec{x}_i) \tag{4.22}$$

In the last step, we introduced a shorthand by building a scalar product of the vector $\vec{p}$ (4.19) with the logarithm of the two normalized network outputs $\vec{q}$ (4.18). An optimally trained network gives $\hat{z}_1 = 1$ for signal events and $\hat{z}_2 = 1$ for background events exclusively. This implies that the objective function becomes $\mathcal{L} = 0$.

Example 4.3. Functionality of the cross-entropy: To understand how the objective function $\mathcal{L}$ in (4.21) drives the optimization process, we take the following example as an intermediate result during the training process. For the signal event $\vec{x}_k$ consider this network prediction:

$$\begin{pmatrix} \hat{z}_1(\vec{x}_k) \\ \hat{z}_2(\vec{x}_k) \end{pmatrix} = \begin{pmatrix} 0.8 \\ 0.2 \end{pmatrix} \tag{4.23}$$

In the objective function the scalar product of the two vectors results in the contribution $\mathcal{L} = (-1) \cdot p_1 \cdot \log(\hat{z}_1) = (-1) \cdot 1 \cdot \log(0.8) \approx 0.1$. Because $p_2 = 0$ for signal events, there is no further contribution as $p_2 \cdot \log(\hat{z}_2) = 0$. In contrast, a signal event $\vec{x}_r$

$$\begin{pmatrix} \hat{z}_1(\vec{x}_r) \\ \hat{z}_2(\vec{x}_r) \end{pmatrix} = \begin{pmatrix} 0.2 \\ 0.8 \end{pmatrix} \tag{4.24}$$

incorrectly classified by the network as a background will result in the significantly larger contribution $\mathcal{L} = 0.7$ to the objective function and motivates further improvements of the network parameters for the network training.

4.5 Gradients from backpropagation

To achieve the minimization of the objective function $\mathcal{L}$ (4.11) or (4.21) and thus the desired network model, we use the so-called *gradient descent* method. It is based on *partial derivatives* of the objective function with respect to the network parameters $\mathbf{W}$ and $\vec{b}$ (3.1).

To obtain these gradients, the backpropagation method is used, which can best be explained using a single network node. In Fig. 4.2, the network prediction $\sigma(W x + b)$ (3.6) and its comparison to the true value y through

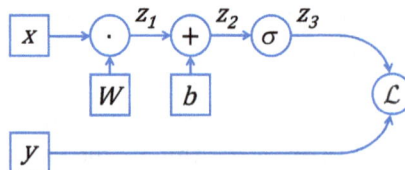

Fig. 4.2. Step-by-step operations of a single network node to explain backpropagation.

the objective function $\mathcal{L}$ is decomposed into small-step operations:

$$z_1 = W \cdot x \tag{4.25}$$
$$z_2 = b + z_1 \tag{4.26}$$
$$z_3 = \sigma(z_2) \tag{4.27}$$
$$\mathcal{L} = (y - z_3)^2 \tag{4.28}$$

It is referred to as a *forward pass* to contrast it with the backpropagation method explained below.

The key question is, how can the parameters W and b be changed to achieve a smaller value of the objective function $\mathcal{L}$? For an efficient procedure, we form the partial derivative of the objective function according to the weight parameter W, which can be separated into individual calculation steps according to the chain rule:

$$\frac{\partial \mathcal{L}}{\partial W} = \frac{\partial \mathcal{L}}{\partial z_3} \cdot \frac{\partial z_3}{\partial z_2} \cdot \frac{\partial z_2}{\partial z_1} \cdot \frac{\partial z_1}{\partial W} \tag{4.29}$$

For the bias parameter b the derivatives are calculated correspondingly:

$$\frac{\partial \mathcal{L}}{\partial b} = \frac{\partial \mathcal{L}}{\partial z_3} \cdot \frac{\partial z_3}{\partial z_2} \cdot \frac{\partial z_2}{\partial b} \tag{4.30}$$

Since the calculation of the derivatives is done in the opposite direction of the forward pass illustrated in Fig. 4.2, this procedure is called *backpropagation*.

4.6 Stochastic gradient descent

Figure 4.3 presents the general idea of searching for a minimum of the target function $\mathcal{L}$ to optimize the network parameters. The gradients, i.e. the partial derivatives $\partial \mathcal{L}/\partial W$ (4.29) and $\partial \mathcal{L}/\partial b$ (4.30), indicate how the parameters W and b of all nodes in all layers need to be changed to minimize the objective function $\mathcal{L}$. Note that these partial derivatives point in the opposite direction in which the parameters are to be changed (see Example 4.5 below).

Example 4.4. Gradient descent: For an exemplary calculation on the node (Fig. 4.2) we will specify the following situation: The current network parameter settings are $W = 0.5$ and $b = 0.1$. As activation function, we use $\sigma = ReLU$ (Fig. 5.8). The point pair (x, y) to be checked has the values $(x = 1, y = 1)$. From these specifications, we get the following value for the objective function, which is chosen to be the mean squared error here (4.11):

$$
\begin{aligned}
z_1 &= & W \cdot x & & = 0.5 \\
z_2 &= & z_1 + b & & = 0.6 \\
z_3 &= & \sigma(z_2) & & = 0.6 \\
\mathcal{L} &= & (z_3 - y)^2 & & = 0.16
\end{aligned}
\tag{4.31}
$$

We apply the chain rule (4.29) to the equations in (4.31). The results of the individual terms are

$$
\begin{aligned}
\frac{\partial z_1}{\partial W} &= & x & & = 1 \\
\frac{\partial z_2}{\partial z_1} & & & & = 1 \\
\frac{\partial z_3}{\partial z_2} &= & \frac{\partial \sigma(z_2)}{\partial z_2} & & = 1 \\
\frac{\partial \mathcal{L}}{\partial z_3} &= & 2(z_3 - y) & & = -0.8
\end{aligned}
\tag{4.32}
$$

When taking the derivative of the $ReLU$ function, we have considered the positive sign of z_2 in (4.31). The partial derivative (4.29) of the objective function about the parameter W thus yields

$$
\frac{\partial \mathcal{L}}{\partial W} = 1 \cdot 1 \cdot 1 \cdot (-0.8) = -0.8 \, .
\tag{4.33}
$$

The derivative $\partial \mathcal{L} / \partial b$ can be calculated accordingly using (4.30).

The gradient of the objective function $\mathcal{L}$ with respect to the weight parameters W is calculated as an average of k training events. The gradient for a bias parameter b is calculated accordingly. The expectation values of the gradients denote:

$$
\mathbb{E}\left[\frac{\partial \mathcal{L}}{\partial W}\right] = \frac{1}{k} \sum_{i=1}^{k} \left(\frac{\partial \mathcal{L}}{\partial W}\right)_i
\tag{4.34}
$$

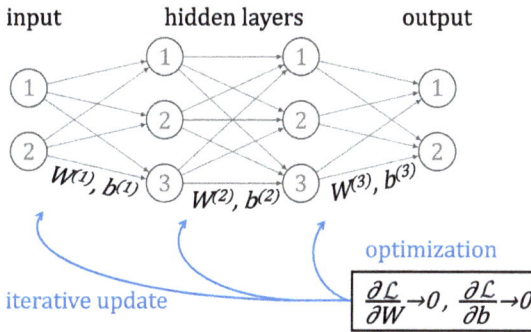

Fig. 4.3. Optimization of the network parameters by iterative search for a minimum of the objective function $\mathcal{L}$ through backpropagation using the partial derivatives $\partial \mathcal{L} / \partial W$ and $\partial \mathcal{L} / \partial b$.

$$\mathbb{E}\left[\frac{\partial \mathcal{L}}{\partial b}\right] = \frac{1}{k} \sum_{i=1}^{k} \left(\frac{\partial \mathcal{L}}{\partial b}\right)_i \qquad (4.35)$$

For an efficient estimation of the gradients, only a minibatch of size k of the training data is used in every optimization step. Using minibatches implies a larger variance of the resulting gradients compared to using the full training data set (one epoch). This enables multiple parameter updates per epoch, and the increased variance sometimes helps to escape from unwanted local minima. Owing to the larger variance introduced by using minibatches instead of the full training data, the procedure is referred to as *stochastic gradient descent* (SGD).

Example 4.5. Parameter update: Figure 4.4 shows an exemplary optimization step t in which the parameter W_t does not yet follow the training data. The opposite direction of the gradient $\partial \mathcal{L} / \partial W$ is then used to calculate for the next step $t + 1$ the parameter W_{t+1}. This lowers the objective function $\mathcal{L}$ and simultaneously gives an improved description of the training data.

4.7 Learning rate

The gradient descent method evaluates for each parameter W by partial derivative of the objective function $\partial \mathcal{L} / \partial W$ whether the parameter W needs

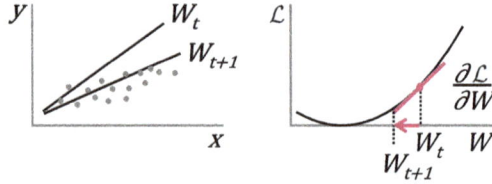

Fig. 4.4. Exemplary optimization step where the negative of the partial derivative $\partial\mathcal{L}/\partial W$ gives the direction for lowering the objective function and improving the description of the training data.

to be increased or decreased in the following iteration step (4.34). To obtain the step size of the parameter change from iteration t to iteration $t+1$, the partial derivative is multiplied by a factor α, which is called the *learning rate*:

$$W_{t+1} = W_t - \alpha\, \mathbb{E}\left[\frac{\partial\mathcal{L}}{\partial W}\right]_t \tag{4.36}$$

The same procedure applies to the bias parameters b (4.35):

$$b_{t+1} = b_t - \alpha\, \mathbb{E}\left[\frac{\partial\mathcal{L}}{\partial b}\right]_t \tag{4.37}$$

The learning rate has a considerable influence on whether, and if so, how fast, a good local minimum of the hypersurface of all parameters can be reached. Figure 4.5 shows that if the learning rate is too low, the minimum may not be reached at all. If the learning rate is too high, a good local minimum could be overlooked. Figure 4.5 also shows that these effects can be observed in the curve of the objective function.

Learning rates are therefore varied in the process of network training. Initially, the starting region on the hypersurface of all parameters is examined with larger step sizes, where learning rates in (4.36), (4.37) are usually in the range:

$$\alpha = 10^{-5} \dots 0.01 \tag{4.38}$$

In the subsequent stages of the training, the learning rate is reduced, and in this way, the hypersurface of parameters is examined in smaller steps.

4.8   Learning strategies

Various strategies exist to accelerate network optimization by steering the learning rate and considering the previous steps in the optimization procedure.

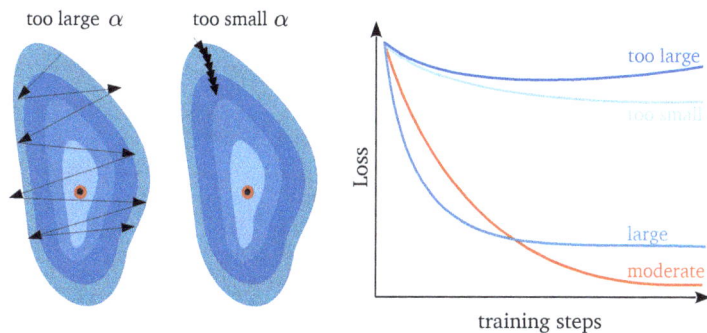

Fig. 4.5. Exemplary learning rates and their impact on the objective function.

One possible approach is to reduce the learning rate whenever no decrease of the objective function $\mathcal{L}$ is observed. Further strategies are presented in the following. None of the methods is generally the best choice, so experimenting with different optimization techniques is recommended.

4.8.1 *Adagrad*

The Adagrad optimizer features *adaptive learning rates*. During optimization the overall learning rate α is continuously reduced, whereby the changes in learning rates are adapted individually for each network parameter [20]. For this, the sum of the squares of all previous gradients is taken into account. For parameters with steep gradients of the objective function, the learning rate should decrease faster than for parameters with flat gradients. Considering, e.g. the weight parameter W, in step t its learning rate α_t is:

$$\nu_t = \sum_{r-1}^{t} \left(\frac{\partial \mathcal{L}}{\partial W} \right)^2 \tag{4.39}$$

$$\alpha_t = \frac{\alpha}{\sqrt{\nu_t} + \varepsilon} \tag{4.40}$$

Here α is the global learning rate (4.38) and $\varepsilon \sim 10^{-8}$ prevents division by zero.

4.8.2 *RMSprob*

The RMSprop optimizer, like the Adagrad algorithm, considers adaptive learning rates [21]. In order to avoid a too rapid decrease of the learning rate, the gradients from far past optimization steps are suppressed by a decay rate parameter with a typical value of $\beta = 0.9$. Considering again the weight parameter W as example, its learning rate α_t at step t denotes:

$$\nu_t = \beta \, \nu_{t-1} + (1 - \beta) \left(\frac{\partial \mathcal{L}}{\partial W} \right)^2 \tag{4.41}$$

$$\alpha_t = \frac{\alpha}{\sqrt{\nu_t} + \varepsilon} \tag{4.42}$$

4.8.3 *Momentum*

Another type of improvement in parameter optimization can be obtained by regarding space points and gradients as vectors in parameter space [22]. Using a single network node as a representative, the two-dimensional vectors are:

$$\vec{\theta} = \begin{pmatrix} W \\ b \end{pmatrix} \qquad \nabla \vec{\theta} = \begin{pmatrix} \frac{\partial \mathcal{L}}{\partial W} \\ \frac{\partial \mathcal{L}}{\partial b} \end{pmatrix} \tag{4.43}$$

The nomenclature momentum reminds of mass times velocity, with the latter being related to the time-derivative of path length. Here the goal is to obtain the most efficient step size in the right direction in parameter space. This is expressed in terms of velocity, or momentum when setting the mass to unity. The velocity corresponding to a standard step from $\vec{\theta}_t$ to $\vec{\theta}_{t+1}$ according to (4.36) and (4.37) reads:

$$\vec{v}_t = -\alpha \, \nabla \vec{\theta}_t \tag{4.44}$$

$$\vec{\theta}_{t+1} = \vec{\theta}_t + \vec{v}_t \tag{4.45}$$

The new aspect of the *momentum* method is to stabilize the direction of optimization by including a history of the velocity [Fig. 4.6(a)]. The direction of the previous velocity $\vec{v}_{t-1}$ is taken and modified by $\nabla \vec{\theta}_t$ to give the velocity $\vec{v}_t$ for the next step:

$$\vec{v}_t = \beta \, \vec{v}_{t-1} + (1 - \beta) \left(-\alpha \nabla \vec{\theta}_t \right) \tag{4.46}$$

$$\vec{\theta}_{t+1} = \vec{\theta}_t + \vec{v}_t \tag{4.47}$$

Here β gives a weight balance between the previous velocity and its modification by the current gradient, down-scaled by the learning rate. Typical values are $\beta = 0.5 \ldots 0.9$.

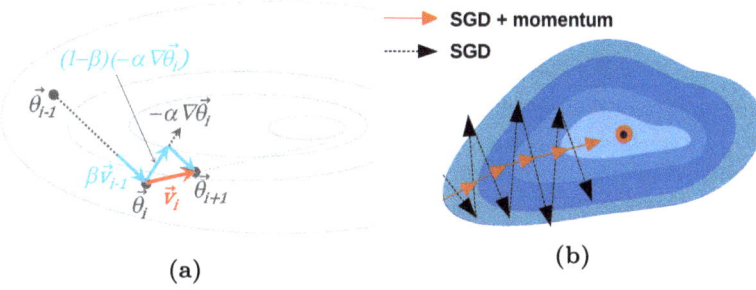

Fig. 4.6. Learning rate momentum, (a) vector sum to calculate next iterative step according to Eq. (4.47), (b) general idea of damping oscillations in the learning process (SGD = stochastic gradient descent).

By roughly maintaining the velocity vector's direction, oscillations on the path through the parameter space can be damped [Fig. 4.6(b)].

4.8.4 *Adam*

The Adam (adaptive moments) optimizer combines the ideas of the RMSprob and momentum-based methods [23]. Both the gradients and the squared gradients are considered with a decay parameter reducing the impact of past values:

$$m_t = \frac{1}{1 - \gamma^t} \left[\gamma \, m_{t-1} + (1 - \gamma) \, \frac{\partial \mathcal{L}}{\partial W} \right] \tag{4.48}$$

$$\nu_t = \frac{1}{1 - \beta^t} \left[\beta \, \nu_{t-1} + (1 - \beta) \left(\frac{\partial \mathcal{L}}{\partial W} \right)^2 \right] \tag{4.49}$$

Proposed values for the decay parameters are $\gamma = 0.9$ and $\beta = 0.999$. The terms preceding the brackets perform initial bias corrections and will lose influence when $t \gg 1$.

The gradient term m_t is used to scale the direction of the next step in parameter space (see momentum method), and the squared gradient term ν_t is used to adapt the learning rate (see RMSprop method):

$$\vec{v}_t = -\alpha \frac{m_t}{\sqrt{\nu_t} + \varepsilon} \tag{4.50}$$

$$\vec{\theta}_t = \vec{\theta}_{t-1} + \vec{v}_t \tag{4.51}$$

As the Adam algorithm and its improvement [24] are fairly robust and efficient, it is a good first choice as optimization algorithm.

4.9   Exercises

Exercise 4.1. We continue with our quest to predict wine quality from physical measurements using neural networks (see Exercise 3.2 for an introduction of this task). The goal is to implement a neural network training with one input layer, one hidden layer, and one output layer using gradient descent. Define the matrices and initialize with random values. We need:

- **W**, with shape: (number of hidden nodes, number of inputs)
- $\vec{b}$, with shape: (number of hidden nodes)
- **W′**, with shape: (number of hidden nodes, 1)
- $\vec{b}'$, with shape: (1)

(1) Use the notebook (Exercise 4.2 from the **exercise page**) to access the data and additional material.
(2) Implement a forward pass of the network.
(3) Implement a function that uses one data point to update the weights using gradient descent. You can follow the provided example.
(4) Now, you can use the provided training loop and evaluation code to train the network for multiple epochs.

Exercise 4.2. Linear Regression: Make yourself familiar with the Keras syntax. From the exercise page, copy the example script (Exercise 4.3) that aims at creating a simple linear model.

Exercise 4.3. Classification: Make yourself familiar with the Keras syntax. From the exercise page, investigate the example script (Exercise 4.4) to learn about one-hot encoding, softmax and cross-entropy. Expand the script:

(1) Implement one-hot encodings
(2) Implement softmax activation
(3) Implement cross-entropy
(4) Evaluate the predicted classes

Summary 4.1.

(1) To achieve numerical stability of the network training, careful preparation of the input data is essential. By *normalizing* the input data, high fluctuations in the data are avoided, and the numerical range is optimally filled for the training process.

(2) For optimization of the network parameters, the available training data is used multiple times. One complete utilization in the training process is called *epoch*. To increase efficiency, parameter optimization is performed in numerous steps using randomly selected subsets of the data, termed *minibatch*, or a bit sloppily *batch*. The number of events or objects in a minibatch is named *batch size*.

(3) For initialization of the weight parameters **W**, random numbers are used to stimulate the training of different mappings in the network nodes. The initialization aims at unit variance in each layer and utilizes a random distribution. The variance of this distribution depends on the number of network nodes in a layer and the activation function. The bias parameters $\vec{b}$ are initialized to zero.

(4) The most common *objective function* (also called *loss* or *cost*) for regression problems is the *mean squared error*, which calculates the average of the square of the residual between network prediction and expected value. For classification problems, the standard objective function is the *cross-entropy* function, which quantifies the level of incorrect classifications.

(5) The gradient descent method utilizes partial derivatives of the objective function for each network parameter with the goal to minimize the objective function. Starting with the chain rule at the network output leads to the term *backpropagation*. All parameters can be improved in each iteration in a targeted manner, which accelerates the optimization process. The gradients are estimated using data subsets (minibatches), hence the term *stochastic gradient descent*.

(6) The *learning rate* corresponds to the step size in the iterative optimization of the network parameters. Different learning strategies exist to search for a good local minimum on the hyperspace of all network parameters. These strategies dynamically adapt the learning rate and the step direction in parameter space during the network training.

Chapter 5

Mastering model building

Scope 5.1. In this chapter, we specify quality criteria for the successful building of network models. We motivate the use of partial data sets for network training, validation, and test of the network and show standard measures for evaluating the network prediction. We introduce regularization methods that stabilize the network training. Hyperparameters such as the number of network layers and nodes specify a network model. The optimal hyperparameter settings are obtained from a scan. In addition, an activation function suitable for the task needs to be selected.

5.1 Criteria for model building

The goal is to obtain the most accurate network predictions for all possible input values. A network's ability to perform well on data that are statistically independent of the training data (but drawn from the same distribution) is referred to as *generalization* capability.

Initially, we will detail criteria for evaluating the network performance. Later in this chapter, we will show measures using the objective function to control the network's generalization performance.

Variance, bias: The optimization of the network parameters is a data-driven process. Thereby, we use a number n of data pairs of real numbers (x_i, y_i), which are taken from a true distribution $g(x)$ by means of $y_i = g(x_i) + \varepsilon$. Here ε is a noise term causing the data y_i to be randomly scattered around $\varepsilon = 0$. The variance of this scattering term ε is named σ^2. The network prediction $f(x)$ is optimized here by minimizing the objective function mean squared error (see regression in Sec. 4.4.1 and (4.10)):

$$\text{MSE} = \mathbb{E}\left[(f - g)^2\right] \tag{5.1}$$

The expectation value of the quadratic deviation between the distribution of the network predictions $f(x)$ and the target distribution $g(x)$ can be divided into three terms according to the bias-variance relation [25]:

$$\text{MSE} = \left(\text{bias}\left[\mathbb{E}[f], \mathbb{E}[g]\right]\right)^2 + V\left[f\right] + \sigma^2 \tag{5.2}$$

The first term *bias* represents the displacement of the expectation value of the network prediction versus the true distribution's expectation value. The second term is the *variance* of the network prediction f, which measures the network quality. The final term denotes the variance σ^2 of the nuisance ε which arises when the value pairs (x_i, y_i) are sampled from the true distribution $g(x)$. It constitutes an *irreducible uncertainty*.

Figure 5.1 shows examples of related generalization problems. In the vertical direction, the bias is varied. In the horizontal direction, the variance is changing. A good generalization is achieved when there are only small shifts and little variance (low bias, low variance).

Problems occur when the network prediction is shifted (high bias, low variance) and when there is a high variance of the network prediction (high variance combined with low or high bias). If a bias in the network prediction can be corrected outside the network (e.g. by a subsequent calibration procedure), a low variance may be the best guiding line when developing the network model.

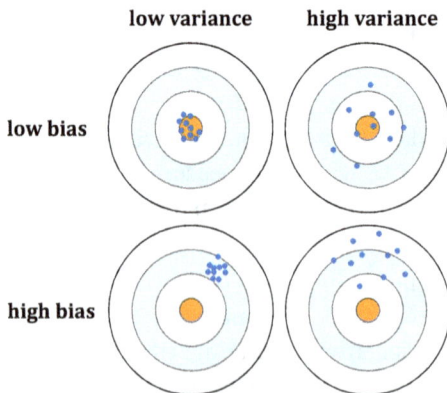

Fig. 5.1. Visualization of bias and variance.

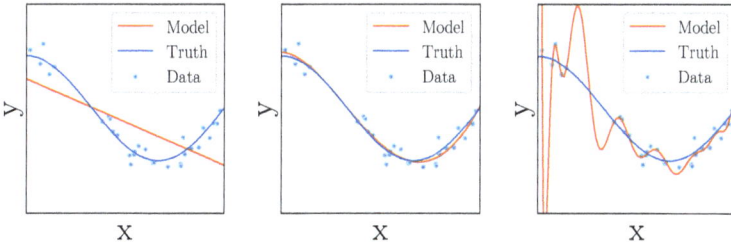

Fig. 5.2. Model qualities: underfitting, appropriate, overfitting.

Overfitting, underfitting: A network model for analyzing data should capture the complexity of the underlying distribution and should not be overly complicated, nor should it be too simple. In the middle of Fig. 5.2, the model chosen is appropriate for the complexity of the data. The data points are scattered around the true distribution which is followed by the network model.

In the left figure, the model is over-simplified, the shape of the data is apparently not captured (*underfitting*). Similarly, in the right figure, the network model follows fluctuations in the data instead of the true distribution (*overfitting*). Usually, neural networks tend to be large models with many parameters, so that overfitting, in particular, will need to be monitored.

Evaluation measures: For regression tasks, bias and variance of the network prediction need to be evaluated as explained above. For classification problems, measures for correct network predictions and false-positive classifications need to be determined. Various definitions are in use, which is discussed here for simplicity in terms of a binary case of an object or event belonging to class A or *not* to class A (i.e. $\bar{A}$).

Accuracy is an overall measure of the correct classification rate. For this ratio the number N_A^{true} of correct assignments to a class A (true positives) and the number $N_{\bar{A}}^{\text{true}}$ of correct other assignments (true negatives) are summed up and divided by the total number N^{total} of classified objects:

$$\text{accuracy} = \frac{N_A^{\text{true}} + N_{\bar{A}}^{\text{true}}}{N^{\text{total}}} \tag{5.3}$$

Sensitivity or *efficiency* is the number N_A^{true} of correct network assignments to class A divided by the total number N_A^{total} of objects truly belonging to class A:

$$\text{sensitivity} = \frac{N_A^{\text{true}}}{N_A^{\text{total}}} \tag{5.4}$$

Precision or *purity* is the number N_A^{true} of correct network assignments to class A divided by the number $N_A^{\text{true}} + N_A^{\text{false}}$ of objects assigned by the network to class A:

$$\text{precision} = \frac{N_A^{\text{true}}}{N_A^{\text{true}} + N_A^{\text{false}}} \tag{5.5}$$

Area under the curve: In classification tasks, one could select the class with the highest probability assignment of a network, i.e. $p_A > 0.5$ in a two-class problem. Alternatively, one can select class A if the network prediction exceeds a threshold value $p_A > v$ with $v \in [0, 1]$, referred to as *working point*. This can be advantageous when deciding on rare processes. Choosing a working point relates the above-described evaluation measures' efficiency and purity: The higher the efficiency for a given threshold v, the lower the purity, and vice versa. The plot of (1-purity) for different efficiency thresholds v is called Receiver Operating Curve (ROC). Its integral, the area under the curve (AOC), describes the overall probability that the network correctly assigns an event from class A to A rather than incorrectly assigning an event from $\bar{A}$ to class A.

Depending on the challenge to be solved by the network, the user decides on acceptable measures.

5.2 Data sets for training, validation, test

Be aware that only a subset of the total available data set should be used for the network training. Otherwise, we will not be able to judge the quality of network training. The whole data set is randomly divided into three parts:

(1) The *training* data set,
(2) The *validation* data set,
(3) The *test* data set.

The exact naming of the data sets (2) and (3) depends on the user community.

Only the training data set (1) is used to train the network. After training, the network prediction quality is checked with the validation data set

(2). Typically, improvements are made to the network afterward, e.g. the number of network nodes in the hidden layers is increased. The network is then retrained. Through this iterative procedure, the validation data set is indirectly involved in the network's training process.

Therefore, it is important to use the last remaining data set, the test data set (3), only after all network design decisions have been made and the network parameters are ready and optimally set. With the test data set, we finally determine the quality of the network prediction once.

The split of the data set can be evenly set to 1/3. Often the total number of objects or events is limited. In this case, priority is given to the optimization of thousands or millions of network parameters, while the proportions of validation and test data sets are reduced. A typical example is to use 60% of the data as a training data set, 15% for the validation data set, and 25% for the test data set. Further advanced procedures exist under the name *cross-validation* [25].

5.3 Monitoring

The objective function (4.11), (4.22) and its evolution during network training give much information on the quality of the network model. Thus, it is continuously evaluated on the training data and carefully monitored as a function of the number of epochs.

To avoid overtraining (Fig. 5.2), also the validation data is repeatedly used to check the quality of the network prediction (Fig. 5.3). At a fixed epoch, the objective function is evaluated separately on the training and validation data. The resulting difference is referred to as *generalization gap*. For networks used for classification tasks, the *accuracy* (5.3) of the

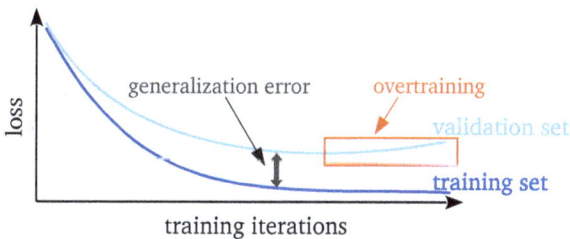

Fig. 5.3. Monitoring the development of the objective function by the validation data to avoid overtraining.

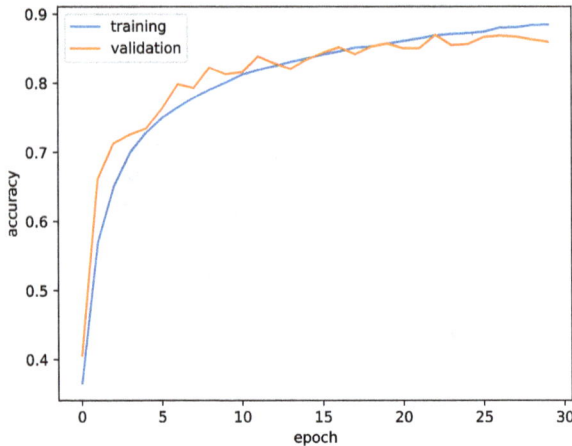

Fig. 5.4. Accuracy as a function of the training epoch.

classification is often evaluated as a function of the epoch in addition to
the objective function (Fig. 5.4).

5.4   Regularization

Network training aims to reach in the hyperspace of a thousand or million
parameters a good local minimum at least approximately and quickly. To
avoid the iterative training getting stuck in an unacceptable local minimum,
several helpful *regularization* methods exist.

According to Ian Goodfellow *et al.*, it is [1]:

> *Regularization is any modification we make to a learning algo-*
> *rithm that is intended to reduce its generalization error but not*
> *its training error.*

The network could learn special properties in the training data that
have no general validity. An example is overtraining shown in Fig. 5.2. To
check for such a bias, the objective function is evaluated on validation data.

Early stopping: Usually, it is assumed that a small bias is not damaging
as long as the objective function is reduced on training data and validation
data. A critical point in training is reached when the objective function
evaluated on the validation data starts rising (Fig. 5.3). This is commonly
interpreted as the beginning of overtraining, such that network training is

terminated at this point. The parameters of the previous epoch are then used.

Parameter norm penalties: As a stabilizing effect for the training process, the objective function can be extended by *penalty* terms, which prevent the formation of numerically large weight parameters **W**. For this purpose, the absolute values of all N weight parameters are summed up before being multiplied by a pre-factor λ and added to the original objective function, e.g. the mean squared error $\mathcal{L}_{\mathrm{MSE}}$ (4.11). This regularization method is called the *L1 norm*:

$$\mathcal{L} = \mathcal{L}_{\mathrm{MSE}} + \lambda \sum_{i=1}^{N} \sum_{j=1}^{N} |W_{ij}| \tag{5.6}$$

Alternatively, the *L2 norm* or *weight decay* is often used, in which the squares of all weight parameters are summed:

$$\mathcal{L} = \mathcal{L}_{\mathrm{MSE}} + \lambda \sum_{i=1}^{N} \sum_{j=1}^{N} W_{ij}^2 \tag{5.7}$$

Figure 5.5 shows the application of the L2 norm as an additional term in the objective function for two weight parameters. Although a local minimum of the mean squared error in the hyperplane of W_1 and W_2 appears in the upper right quadrant, the L2 term in the overall objective function motivates the reduction of weights, especially the numerically large value W_2 which itself has a shallow gradient. This is in contrast to the L1 norm. As visible in Fig. 5.5(a), L1 regularization pushes unimportant

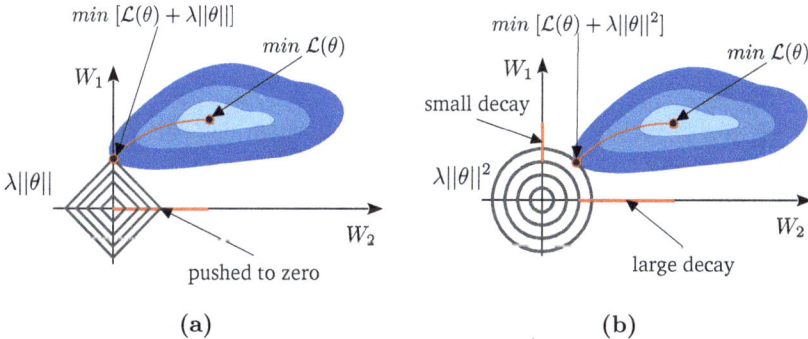

Fig. 5.5. Parameter norm penalties: (a) L1 norm, (b) L2 norm.

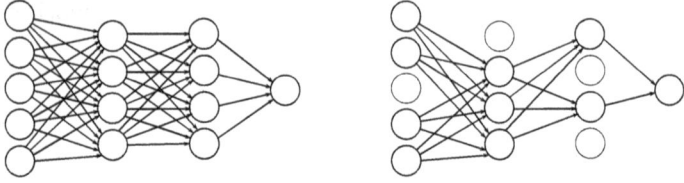

Fig. 5.6. Regularization by dropout.

parameters (here W_2) to zero. Hence, using parameter norm penalties a new local minimum in the optimization process is encouraged, which yields a more balanced weight distribution and possibly better generalization performance.

Dropout: The dropout method's name is programmatic (Fig. 5.6): During network training, individual network nodes are cut off at random. Typical dropout rates, specifying which fraction of nodes to drop, vary between $0.2 \ldots 0.5$. This stimulates the network to create several mappings for input values and their expected network prediction while optimizing the weight parameters. Often this method improves the stability of the network. Dropout is only used during training. When using the network on independent data for predictions, all network nodes are used. To preserve the overall magnitude of the network layer output despite missing nodes, during the training phase the weights of the active nodes are scaled inversely with the dropout rate.

Further regularization measures: There are other helpful measures to stabilize the network training. Especially the use of considerably more training data is a common and very successful means. An efficient method is to exploit symmetries in existing training data, e.g. by translation or rotation operations, and in this way use the same (modified) data multiple times. This is referred to as *data augmentation*.

Another method alters the input values during training by adding noise. This measure also has a stabilizing effect on the training. This is known as *noise injection*.

Furthermore, several networks can be trained for the same purpose, and the predictions of all networks can be combined for a more reliable prediction. This is referred to as *ensemble training*.

Finally, network parameters can be correlated with each other, or the same parameter value can be used in several places. These methods are termed *parameter tying* or *parameter sharing*. Later, we will see that such methods are used in certain architecture concepts, such as in convolutional, recurrent, or graph networks (see Part 2).

5.5 Hyperparameters and activation function

Hyperparameters: For the construction of a network model, many decisions have to be made. Typically these include

(1) the number of hidden layers (Fig. 1.3),
(2) the number of network nodes in a layer (Fig. 1.3),
(3) the choice of the activation function (Fig. 5.8),
(4) the way the network parameters should be initialized (4.8), (4.9)
(5) the coefficients of the regularization terms (5.6), (5.7),
(6) the batch size (Sec. 4.2),
(7) the learning rate (4.36), (4.37), and
(8) the learning strategy (Sec. 4.8).

To distinguish these basic parameter settings of the network model from the weight parameters $\mathbf{W}$ and bias parameters $\vec{b}$, they are called *hyperparameters*.

It is difficult to predict in advance which values of a given hyperparameter should be selected. Therefore, it is recommended to perform a scan of the hyperparameters. The objective function is used as a measure for the best settings. The scan can be performed on a regular grid (*grid search*). However, since several hyperparameters are of different importance it is advisable to optimize such a scan by randomly selecting the hyperparameters (*random search*) or using more advanced statistical methods. Figure 5.7 illustrates the advantage of using a random hyperparameter scan.

Activation function: A further decision needs to be taken on the *activation function* (Sec. 3.2) which is chosen depending on the desired modeling. We present some basic examples in Fig. 5.8. The ReLU function turns out to be a good first choice [Fig. 5.8(a)]. The leaky ReLU function is a direct extension in that negative results can also be passed on [Fig. 5.8(b)] [26].

The sigmoid [Fig. 5.8(c)] was one of the first introduced activation functions due to its similarity to neuroscience. The sigmoid is rarely used

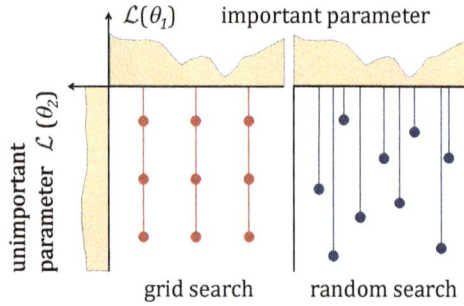

Fig. 5.7. Hyperparameter search: left grid search, right random search. Using random search, various more values of θ_1, θ_2 can be assessed.

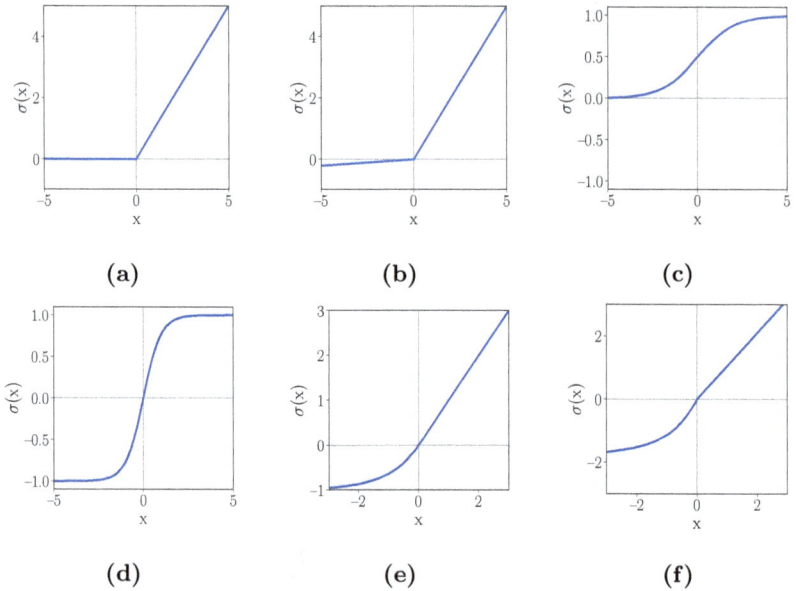

Fig. 5.8. Examples of activation functions. (a) ReLU, (b) leaky ReLU, (c) sigmoid, (d) hyperbolic tangent, (e) ELU, and (f) SELU.

recently in-between layers as the mean of the produced activation is always larger than zero, potentially leading to disturbances in deep networks. Additionally, the saturation regimes of the sigmoid and the hyperbolic tangent [Fig. 5.8(d)] can lead to vanishing gradients and thus a stopping of the training. Therefore, the sigmoid and hyperbolic tangents are mostly used as final layers of a model to constrain the prediction range to $[0, 1]$ or rather $[-1; 1]$. We discuss this so-called *vanishing gradient problem* in-depth in Sec. 7.4.

For the more advanced activation functions ELU [Fig. 5.8(e)] and SELU [Fig. 5.8(f)], we refer to our discussion in Sec. 7.7.

5.6   Exercises

Exercise 5.1. Regularization: Open the Tensorflow Playground (`playground.tensorflow.org`) and select on the left the checkerboard pattern as the *data* basis (see Example 3.3). In *features*, select the two independent variables x_1 and x_2 and set noise to 50%. Start the network training and describe your observation when doing the following steps:

(1) Choose a deep and wide network and run for > 1000 epochs.
(2) Apply L2 regularization to reduce overfitting. Try low and high regularization rates.
(3) Compare the effects of L1 and L2 regularizations.

Note: You can inspect the weights by hovering over the connecting lines.

Exercise 5.2. Interpolation: For a first look at neural network design, training, and verification, follow the main steps of the Keras syntax using the program (Exercise 5.2) from the exercise page:

(1) *Define the network model*: add network layers with their nodes, decide on an activation function, and possibly a regularization procedure.
(2) *Compile the model*: choose the objective function (loss) and a learning strategy.
(3) *Train the model*: Specify the number of iterations for the optimization of the network parameters, and train the network using the training data.

Monitor the evolution of the objective function. After network training is complete, verify the model by inputting test data and observing the network predictions.

Exercise 5.3. Let us visit the problem of wine quality prediction previously encountered in Exercises 3.2 and 4.1 one final time. After linear regression and a self-made network, we can now explore the comfort provided by the KERAS library.

(1) Use the notebook (Exercise 5.3 from the exercise page) as starting point. It already contains the simple network from Exercise 4.1 implemented in KERAS.

(2) Currently, SGD is used without momentum. Try training with a momentum term (see the documentation of SGD). Replace SGD with the Adam optimizer and train using that.

(3) Add two more hidden layers to the network (you can choose the number of nodes but make sure to apply the ReLu activation function after each) and train again.

(4) (more difficult) Instead of single examples, switch to training using batches of five examples. Note that the network can be applied to a matrix of multiple examples. In this case, it will return a vector with one output per example.

Summary 5.1.

(1) *Generalization* means that a network should provide correct predictions for data that have not been used to optimize network parameters.

(2) The *bias-variance relation* shows that through the data-driven optimization of the network parameters, a trade-off between both bias and variance of the network prediction is to be found. Additionally, an irreducible uncertainty from the scatter of the target values remains.

(3) The complexity of a network model should correspond to the research question. Neither should there be *overfitting*, where network prediction tracks the training data but does not generalize, nor should there be *underfitting*, where network prediction neglects the underlying true distribution complexity.

(4) To both train and evaluate the network with data, three partial data sets are extracted from the entire available data: *Training* data and *validation* data enter the training process. The *test* data are used only once for statistically independent data evaluation and final assessment of the network quality.

(5) To evaluate the progress in network training, the evolution of the objective function is *monitored* as a function of the training epoch. The objective function is followed for both the training data and the validation data to monitor the network's generalization capability. For classification problems, the *accuracy* as a function of the epoch can also be checked.

(6) A wide range of *regularization* methods helps to stabilize network training. These include early stopping, parameter norm penalties, dropout, data augmentation, noise injection, ensemble training, parameter tying, and parameter sharing.

(7) *Hyperparameters* are all parameters that specify the model and its training of the model. These include the number of hidden layers, number of nodes, activation function, initialization of network parameters, batch size, coefficients of regularization terms, learning rate and learning rate schedule, and many more. These hyperparameters can be optimized through a scan.

PART 2
Standard Architectures of Deep Networks

In this part, we show variants of networks tailored to specific properties of the input data. Such properties are, for example, translation, rotation, or permutation symmetries. The consideration of such properties in the architecture frequently leads to a performance increase in the networks.

Chapter 6

Revisiting the terminology

We start with a short review of Part 1, focusing on terms relevant for understanding the following chapters on modern deep learning concepts.

A neural network is a model for the mathematical description of, in our case, physics processes (Fig. 1.3). The network parameters (*weight, bias*) are adapted to a specific research question through data (Secs. 2.3 and 3.1). The network output is referred to as *prediction* (Sec. 3.3). Typical learning tasks are *classifications* (discrete targets) or *regressions* (continuous targets) (Secs. 4.4.1 and 4.4.2).

The basic building block of a network is the *layer* with *nodes*, at each of which two operations are performed (*affine mapping, nonlinear activation function*) (Secs. 3.1, 3.2 and 5.5). Specifications of a network (nodes per layer etc.) are summarized by *hyperparameters* (Sec. 5.5). We often refer to variables on input or as results of mappings within the network as *features*. With an increasing number of layers, the features become more complex and *abstract* (*feature hierarchy*). The phase space of network variables is called feature space. In contrast, measurement data is referred to be in the *spatial domain*.

Learning describes the process where a network adopts solving a particular task. *Training* concretely means data-driven tuning of all parameters (θ = weights, biases) of the network model by minimizing an *objective function* $\mathcal{L}$ (*loss, cost*) (Sec. 4.4). All components of the network should be differentiable so that an individual adaptation of each parameter θ_i can be achieved by the gradient $\partial \mathcal{L}/\partial \theta_i$ (*backpropagation, stochastic gradient descent*) (Secs. 4.5 and 4.6). Parameter adjustment belongs to the ultimate challenges in modeling with neural networks and has to be carefully monitored. A catalog of support methods is available (*learning strategies, regularization*) (Secs. 4.8 and 5.4).

Data in physics usually consist of measurements as a function of time at different locations and are represented as sequences, images, vectors, or heterogeneous sets of variables (*sequence, image, point clouds, other*) (Sec. 2.1). Before entering measurement data into the network, suitable normalization of the data is required (*preprocessing*) (Sec. 4.1). So-called *labeled data* contain a target value (*label* = true value) for the network prediction in addition to the input data of the network. Learning with target values (e.g. using simulated data) is called *supervised training*.

To obtain statistically correct results with the network and avoid that the network memorizes the data, the data set is divided into three parts (*training, validation, test*) (Sec. 5.2). The multiple uses of the *training data* set for parameter optimization are measured in units of *epochs* (Sec. 4.2). Training steps are performed with small sub-datasets (*minibatch*), the success is monitored with the *validation data* (Sec. 5.3). Final statistical evaluations of data with the network model are performed with the *test data*.

In the following, we extend our fully-connected network from Part 1 to several layers and later present modern alternative concepts for different types of measurement data. Which of these standard concepts should be used depends on the structure of the measurement data and the scientific question to be addressed.

Chapter 7

Fully-connected networks: improving the classic all-rounder

Scope 7.1. A fully-connected neural network (FCN) represents a basic network architecture. Its components were presented in detail in Chapter 3. Here, we extend these networks to several layers and thus start with *deep learning*. We give details for two concrete applications, a regression, and a classification task. Both problems can be solved with almost the same network architecture. Differences arise in the post-processing of the network outputs, and in the objective functions for optimizing the network parameters. Furthermore, we present three advanced techniques for successful training of deep networks that address the network architecture, the normalization of data, or the activation function: residual networks, batch normalization, and self-normalizing networks.

7.1 N-layer fully-connected networks

Fully-connected layers are the building blocks of traditional neural networks. Each node of such a layer is connected to each node of the following layer (Fig. 7.1). At each node, the same process occurs: linear mapping and bias with adjustable free parameters $\mathbf{W}$ and $\vec{b}$ (3.1). Then the non-linear activation function is applied (3.4). Adding many of these layers consecutively, we exploit the machine learning method referred to as *deep learning*.

In the following, we present two examples of fully-connected networks with several layers. The first one is representative of a regression task. It involves an ensemble of time measurements after an event, which is used to determine the event's original location. The second example shows a typical classification task. Here, images are to be sorted into two categories.

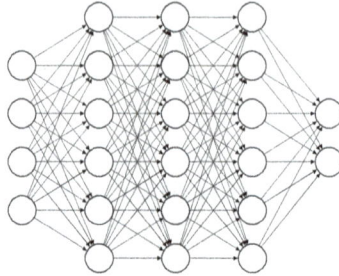

Fig. 7.1. Fully-connected neural network [19].

After these applications, we discuss challenges in training deep networks and different methods as possible solutions.

7.2 Regression challenge: two-dimensional location

The first task is to find the epicenter of earthquakes (see Sec. 2.2). Figure 7.2 shows 1,000 simulated earthquakes occurring along the border of two tectonic plates (black symbols). Four monitoring stations (red symbols) are installed to report the arrival times of seismic signals. Each station reports two time measurements: the arrival time t_p of the compression wave and the arrival time t_s of the shear wave.

Table 7.1 shows the available data. The first column shows the earthquake number. The second and third columns show the true location of the simulated epicenter. This information is the *data label* needed for network training. Note that the label is never given to the network as an input value so that the network calculates the epicenter exclusively from the time information of the measuring stations. Columns four and five

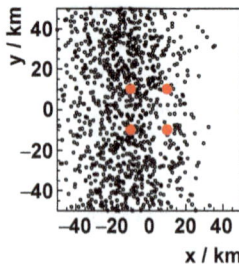

Fig. 7.2. Epicenters of earthquakes (black), measurement stations (red).

Table 7.1. Simulated data of earthquakes with the epicenter (x, y) and four seismic stations each reporting arrival times of compression and shear waves.

i	x/km	y/km	$t_p^{(1)}/$s	$t_s^{(1)}/$s	$t_p^{(2)}/$s	$t_s^{(2)}/$s	$t_p^{(3)}/$s	$t_s^{(3)}/$s	$t_p^{(4)}/$s	$t_s^{(4)}/$s
1	-15.5	-34.6	6.52	9.00	7.25	10.32	3.39	4.98	5.06	7.06
2	3.8	-7.7	3.31	4.51	2.59	3.79	1.79	2.75	0.95	1.30
3	-0.2	29.1	3.17	4.31	3.00	4.37	5.54	8.00	5.77	8.06
4	-16.4	-25.4	5.23	7.21	6.22	8.87	2.17	3.27	4.37	6.09
...	...									

give the measurement times $t_p^{(1)}$ and $t_s^{(1)}$ of the first station, followed by the measurement times of the other three stations. Because of the higher propagation speed of the compression wave compared to the shear wave, the time t_p is always earlier than t_s. Only these arrival times are given to the inputs of the network.

Before inputting to the network, all time measurements are normalized using the largest time in the data set.

The architecture of the network consists of eight inputs for the time measurements $\vec{t}^T = (t_p^{(1)} \, t_s^{(1)} \, t_p^{(2)} \, t_s^{(2)} ...)$ of the measuring stations, which are always entered in the same order. We use $k = 3$ hidden layers with $m = 64$ nodes each. In the first network layer, these time measurements are linearly combined and shifted. Then the activation function σ is applied, where we use the ReLU function. The data processing continues as follows:

$$\vec{z}_0 = \sigma(\mathbf{W}_0 \, \vec{t} + \vec{b}_0) \tag{7.1}$$

$$\vec{z}_1 = \sigma(\mathbf{W}_1 \, \vec{z}_0 + \vec{b}_1) \tag{7.2}$$

$$\vec{z}_2 = \sigma(\mathbf{W}_2 \, \vec{z}_1 + \vec{b}_2) \tag{7.3}$$

In the final step, all results are combined into a two-dimensional result vector $\vec{z}$ with the two values $\hat{x}$ and $\hat{y}$. This vector contains the prediction of the network for the position of the earthquake epicenter, calculated from the input measurement times:

$$\vec{z} = \begin{pmatrix} \hat{x} \\ \hat{y} \end{pmatrix} = \mathbf{W}_3 \, \vec{z}_2(\vec{z}_1(\vec{z}_0(\vec{t}))). \tag{7.4}$$

Here $\mathbf{W}_3$ denotes a $2 \times m$ matrix to combine all nodes of the penultimate layer to the two output values $\hat{x}$ and $\hat{y}$. Note that no activation function σ is used in this last step.

To correctly adjust the adaptive parameters $\mathbf{W}$ and $\vec{b}$ of the network for this calculation, we use the mean squared error (MSE) as the objective function (4.11). It is calculated from the sum of the squared residuals of the true position of the epicenter x_i, y_i (data label) and the network prediction $\hat{x}_i, \hat{y}_i$ for measurement i:

$$\mathcal{L} = \frac{1}{n} \sum_{i=1}^{n} \left[(x_i - \hat{x}_i)^2 + (y_i - \hat{y}_i)^2 \right] \tag{7.5}$$

For the calculation of the objective function only a subset of the data (mini-batch) is used, not the whole dataset. As batch size, we use $n = 32$ here. We select the initial weight parameters $\mathbf{W}$ using a uniform random distribution with the standard deviation $\sigma = \sqrt{6/(n_{\mathrm{in}} + n_{\mathrm{out}})}$, where n_{in} is the number of nodes of the previous layer and n_{out} that of the following layer (Sec. 4.3). The bias parameter $\vec{b}$ is initially set to 0.

During network training with 19,000 earthquakes, the sum of the squared residuals in the objective function is minimized. Thereby, the network predictions for the position of the epicenter steadily improve. This can be followed by the progress of the objective function as a function of the number of epochs both for the training data and for the validation data using another set of 1,000 earthquakes [Fig. 7.3(a)]. The mean distance of the network prediction from the epicenter's true simulated value is only $\Delta r \sim 70$ m [Fig. 7.3(b)] after the training concluded.

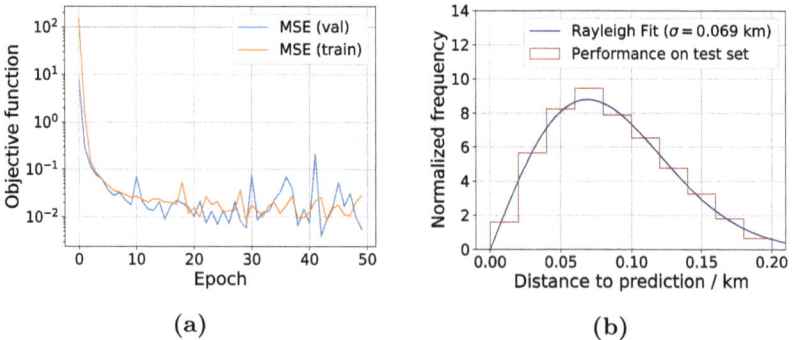

(a)   (b)

Fig. 7.3. (a) Network training and (b) spatial resolution of the epicenter [27].

7.3 Classification challenge: two image categories

As second example for applying a fully-connected network, we take as a thought experiment black and white images of spiral galaxies (Fig. 7.4). Of interest is the rotational orientation of the spirals. The images consist of 32×32 pixels with an intensity value between 0 and 1, where black corresponds to the intensity value $I = 0$ and white to $I = 1$.

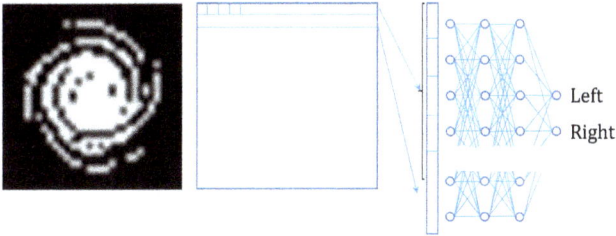

Fig. 7.4. Image recognition with a fully-connected network. The image of a spiral galaxy — here reduced to 32×32 pixels — is processed as a whole. The intensities of the pixels are filled line by line into the 1,024 input values of the network and then processed.

The 32 lines of each image i with their 32 pixels each are arranged one after another in a vector $\vec{x}_i$ with a total of $n = 32 \times 32 = 1,024$ slots. The vector of each image is given to the 1,024 input nodes of a network.[1]

The architecture of the network is almost the same as for the above earthquake network. The main changes are the number n of inputs and the output. Instead of a prediction of a continuously distributed localization of epicenters, a binary orientation specification for galaxies is now required. One possible way is to use only one output scalar z, which, by applying the sigmoid function to z [Fig. 3.3(b)], allows a binary prediction through $z > 0.5$ and $z < 0.5$.

In the following, we show another solution that can also be used for $k \geq 2$ classes. For this purpose two output nodes are used and connected in such a way that the resulting values z_1 and z_2 are normalized to the value 1 and each individual output value can be interpreted as the probability of

---

[1] We discuss a more sophisticated technique for image-like data in Sec. 8.1. The vector representation is not optimal as we ignore symmetry and locality properties of the image data.

the image belonging to this class:

$$\vec{z} = \begin{pmatrix} \hat{z}_1 \\ \hat{z}_2 \end{pmatrix} = \frac{1}{e^{z_1} + e^{z_2}} \begin{pmatrix} e^{z_1} \\ e^{z_2} \end{pmatrix} \tag{7.6}$$

This function formed by exponential functions is called *softmax* and ensures $\hat{z}_1 + \hat{z}_2 = 1$.

For network training, we need a suitable objective function for discrete problems, for which we use the cross-entropy (4.22). For the images used for the network training, we need the true orientation of the galaxy. For each of the $k = 2$ classes, we reserve a component of a k-dimensional vector $\vec{y}$ and fill in a 1 at the position if the image belongs to this class. All other components of the vector are filled with 0. For left or right-oriented spiral galaxies, the vector looks like this:

$$\vec{y}_j^T = \begin{pmatrix} y_1 \ y_2 \end{pmatrix} = \begin{cases} \begin{pmatrix} 1 \ 0 \end{pmatrix} : \text{Right} \\ \begin{pmatrix} 0 \ 1 \end{pmatrix} : \text{Left} \end{cases} \tag{7.7}$$

We refer to this procedure as *one-hot-encoding*.

The normalized network prediction $\vec{z}$ (7.6) is then evaluated by the objective function of the cross-entropy (4.22):

$$\mathcal{L} = -\frac{1}{m} \sum_{i=1}^{m} \begin{pmatrix} y_{i,1} \ y_{i,2} \end{pmatrix} \begin{pmatrix} \log\left[\hat{z}_1(\vec{x}_i)\right] \\ \log\left[\hat{z}_2(\vec{x}_i)\right] \end{pmatrix} \tag{7.8}$$

Here m is the number of images in a minibatch, $y_{i,1}$ ($y_{i,2}$) is the true orientation of image i, $\hat{z}_1$ the probability that the image belongs to galaxies with the spiral orientation to the right, and $\hat{z}_2$ the corresponding probability to belong to left-oriented galaxies.

How network training works by minimizing cross-entropy was explained in Sec. 4.4.2. When training with right or left oriented galaxy images, the network's adaptive parameters are set by minimizing the objective function $\mathcal{L}$ so that the orientations of galaxies can be identified.

7.4   Challenges of training deep networks

Next, we discuss typical challenges in training deep networks and subsequently present possible solutions. With an increasing number of network layers, more input data attributes can be processed to form the network prediction. However, as the network depth increases, also the number of network parameters increases, and so does the challenge of network training. A particular concern is that the gradients (4.34), (4.35) may either *vanish* or *explode* during training deep networks. Both of these effects prevent training progress and thus cancel the desired improvement in more detailed network mappings.

In network training, we optimize a parameter W through the gradient of the objective function $\mathcal{L}$. In Sec. 4.5, we had exemplarily demonstrated the computation of such a gradient $\partial\mathcal{L}/\partial W$ with the backpropagation method on a single network node. For networks with many layers, the gradient calculation for the parameter W_{ij} of node i from layer j involves parameters $W_{k\ell}$ from other layers ($k \neq i, j \neq \ell$). Owing to this interconnection of gradients and weights from different layers, choosing appropriate learning rates α becomes much more challenging. Here, the adaptive learning methods with parameter-wise α_i presented in Sec. 4.8 provide suitable concepts.

An even greater challenge arises with parameter updates. Imagine that in the next iteration of the network training, a parameter W_{ij} of node i from layer j is changed according to its gradient. The parameter update was calculated based on the previous values of parameters $W_{k\ell}$, but they are also changed by their own gradients in the same iteration (see Example 7.1). Coordinating all parameter updates is usually not performed owing to computational time constraints, resulting in unplanned shifts in the network activations. Such changes in the distributions of network activations induced by parameter optimization are called *internal covariate shifts* [28].

Our Example 7.1 of a multi-layer linear network is, of course, an oversimplification of some aspects. Using a nonlinear activation function σ with a saturation region, such as the sigmoid function, vanishing gradients can occur naturally in a network with n layers. Since the value of the sigmoid function is always smaller than 1 for finite input, calculating the nth power of σ may lead to vanishing values: $\sigma^n \to 0$.

There are various ways to stabilize the training process against vanishing or exploding gradients. We present three concepts that address the network architecture, the normalization of data, or the activation function: (a) For the training of *residuals* and improved gradient propagation, the *network architecture* is modified by so-called *shortcuts*. (b) In *batch normalization*, the *minibatch* data are transformed before input to the next network layer. (c) Within *self-normalizing neural networks*, the *activation functions* control the distributions of the activations and cause the activation averages and variances to converge to fixed points.

Example 7.1. Parameter update in multi-layer network: We study a 'three-layer network' with only one node each as an example, with which a linear mapping is performed (activation function $\sigma(x) = x$, bias $b = 0$). We denote the gradients of the parameters by g_i and the learning rate by α. The current network prediction $\hat{z}$ at iteration m for input x reads:

$$\hat{z}^{[m]} = x\, W_1^{[m]}\, W_2^{[m]}\, W_3^{[m]}$$

As the objective function we use the mean squared error (MSE) comparing to the true value $z(x)$:

$$\mathcal{L} = \left[z(x) - \hat{z}^{[m]}(x) \right]^2 \tag{7.9}$$

The gradient of the parameter W_1 reads:

$$g_1 = \frac{\partial \mathcal{L}}{\partial W_1}$$

$$= 2 \left[z(x) - x\, W_1^{[m]}\, W_2^{[m]}\, W_3^{[m]} \right] W_2^{[m]}\, W_3^{[m]}$$

Note that g_1 depends on the actual values of the parameters $W_2^{[m]}\, W_3^{[m]}$ in the other layers, so do the gradients g_2, g_3 of the other two parameters $W_2^{[m]}, W_3^{[m]}$. Depending on their numerical values, the gradients may tend to zero or infinity in the worst case (*vanishing* or *exploding gradients*), preventing successful network training.

For the following iteration $m+1$, all parameters are updated according to their gradients g_i. Thereby, we assume that for a better approximation of the true distribution of $z(x)$, the network's overall update can be factorized into independent updates of the individual parameters.

The factorization assumption leads to two challenges. One challenge is to choose an appropriate learning rate α. The second challenge is the *internal covariate shifts* mentioned above.

Here, we motivate the challenge for choosing an appropriate value for the learning rate α. For this, we expand the network prediction of iteration $m + 1$ in terms of a Taylor series in α:

$$
\hat{z}^{[m+1]} = x\, W_1^{[m+1]} \, W_2^{[m+1]} \, W_3^{[m+1]}
$$

$$
= x\, (W_1^{[m]} - \alpha\, g_1^{[m]})\, (W_2^{[m]} - \alpha\, g_2^{[m]})\, (W_3^{[m]} - \alpha\, g_3^{[m]})
$$

$$
= \underbrace{x\, W_1^{[m]} \, W_2^{[m]} \, W_3^{[m]}}_{=\hat{z}^{[m]}} - x\,\alpha\, \underbrace{g_1^{[m]} \, W_2^{[m]} \, W_3^{[m]}}_{\equiv \tilde{g}_1^{[m]}}
$$

$$
- x\,\alpha\, g_2^{[m]} \, W_1^{[m]} \, W_3^{[m]} - x\,\alpha\, g_3^{[m]} \, W_1^{[m]} \, W_2^{[m]}
$$

$$
+ x\,\alpha^2\, g_1^{[m]} \, (g_2^{[m]} \, W_3^{[m]} + g_3^{[m]} \, W_2^{[m]}) + \ldots \tag{7.10}
$$

Owing to the product terms such as $\tilde{g}_1^{[m]} = g_1^{[m]} \, W_2^{[m]} \, W_3^{[m]}$ the choice of an appropriate learning rate α is challenging as it depends on the parameters of the deeper layers. The parameters and their gradients are all interconnected.

7.5   Residual neural networks

We motivate the learning of residuals by considering the notion of hierarchical learning. In the first layers of a network, the aim is to learn the rough structure of the overall desired mapping $f(\vec{x})$. In the deeper network layers, the goal is to learn some detailed modifications for the overall mapping. Thus only small modifications are to be learned in the deep layers, in the extreme case, merely the identity mapping.

Instead of relying on the network to learn the identity $\mathbb{1}$, it appears more convenient to learn small modifications $\delta(\vec{x})$ to the identity mapping $\mathbb{1}$ [29]. The desired mapping f is therefore composed of δ and $\mathbb{1}$:

$$
f(\vec{x}) = \delta(\vec{x}) + \mathbb{1}\,\vec{x} \tag{7.11}
$$

The nomenclature as *residual* learning becomes obvious when we invert the equation: $\delta = f - \mathbb{1}$.

Technically, the identity mapping is implemented as a so-called *shortcut* [30–32]. As shown in Fig. 7.5, the input $\vec{x}$ to a series of network layers is element-wise summed with the output $\delta(\vec{x})$ of the network layers. For the mapping $\delta(\vec{x})$ from two network layers needed at least, we obtain the following result:

$$
z = \sigma\left(\delta(\mathbf{W}, \vec{b}; \vec{x}) + \vec{x}\right) \tag{7.12}
$$

$$
= \sigma\left(\mathbf{W}_2 \left[\sigma\left(\mathbf{W}_1\,\vec{x} + \vec{b}_1\right)\right] + \vec{b}_2 + \vec{x}\right) \tag{7.13}
$$

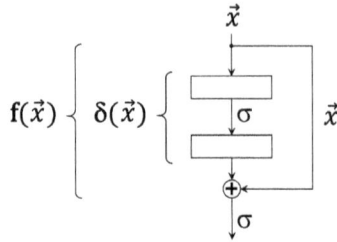

Fig. 7.5. Shortcut of residual learning.

The second activation function σ is applied only after completing the short-cut. Obviously, with such a shortcut, the number of network parameters remains unchanged, and the additional computational effort required by element-wise summation in the forward path is negligible. Nevertheless, for backpropagation during training, the two paths have to be followed.

If the dimensions of $\vec{x}$ and $\delta(\vec{x})$ are different, a linear projection with an appropriate mapping $\mathbf{W}_s$ is recommended:

$$f(\vec{x}) = \delta(\vec{x}) + \mathbf{W}_s\,\vec{x} \tag{7.14}$$

The simplification of the learning process using shortcuts has been successfully used in computer vision (ResNet [29, 33]). There is also ex-perimentation with even more extreme use of shortcuts. Here, e.g. the outputs of each layer are simultaneously used as inputs of the following layers (DenseNet, see Sec. 8.4.1).

7.6   Batch normalization

The so-called *batch normalization* approach is a re-parameterization method applied between network layers [28]. The outputs y of each node j of the previous network layer (usually before applying the activation function σ) are collected for a minibatch of size k to calculate a transformation for each node output according to their mean value μ_B and standard deviation σ_B observed in that batch:

$$\mu_B = \frac{1}{k} \sum_{i=1}^{k} y_i \tag{7.15}$$

$$\sigma_B = \sqrt{\epsilon + \frac{1}{k} \sum_{i=1}^{k} (y_i - \mu_B)^2} \tag{7.16}$$

$$y_i' = \frac{y_i - \mu_B}{\sigma_B} \tag{7.17}$$

Both parameters μ_B and σ_B influence the gradients for the following iteration as they take part in the backpropagation. The parameter ϵ prevents division by zero.

To enable identical mappings, two new network parameters γ_j and β_j are additionally introduced for each activation to re-enable dispersion (γ_j) and shift (β_j):

$$y_j' = \gamma_j \frac{y_j - \mu_B}{\sigma_B} + \beta_j \tag{7.18}$$

This measure is initially surprising as it looks like cancellation of the intended procedure. However, the learning behavior for these new parameters is different since it performs targeted changes of the variance and average, which are not easily accessible in the traditional network design. Experiments show that batch normalization accelerates the training of deep networks and is commonly regarded as a fundamental advance in neural network optimization by ensuring moderate values and gradients. Also, the choice of the learning rate becomes less critical.

7.7   Self-normalizing neural networks

Stabilization of network training can also be achieved via the properties of the activation function [34]. By suitable design, the distribution of the activations of each network layer is deliberately limited.

Such engineered activation functions are deployed in so-called *self-normalizing networks* [35]. They are defined by requiring that the first two moments (mean, variance) of the activation distribution of a network layer can be mapped onto the corresponding moments of the next layer. Besides, the moments need to converge towards fixed points in the iterative training of the network.

We denote the average values and variances of the activations x_i, i.e. the outputs of the activation functions, of a network layer j with n nodes by:

$$\mu_j = \frac{1}{n} \sum_{i=1}^{n} x_i$$

$$\nu_j = \frac{1}{n} \sum_{i=1}^{n} (x_i - \mu_j)^2$$

We write the mapping K of these moments from layer j to layer $j+1$ as:

$$\begin{pmatrix} \mu_{j+1} \\ \nu_{j+1} \end{pmatrix} = K \begin{pmatrix} \mu_j \\ \nu_j \end{pmatrix} \tag{7.19}$$

To get an idea of the mapping K, we first specify the affine mapping $y_i = \sum W_{jk} x_k + b_i$ at each node i of layer $j+1$. For simplicity, we consider the input variables x_k and the parameters W_{jk}, b_i as independent random variables, each taken from appropriate probability distributions. For a sufficiently large number of nodes in layer $j+1$, the affine mappings y_i jointly follow a Gaussian distribution according to the central limit theorem.

In the two-step computation at each network node, the affine mapping y_i is followed by the activation function σ. Convolutions of the Gaussian probability density p_G of the y_i with σ yield the moments of the activation distribution after network layer $j+1$. The mean and variance of the mapping K are obtained from:

$$\mu_{j+1} = \int_{-\infty}^{\infty} \sigma(y) \, p_G(y) \, dy$$

$$\nu_{j+1} = \int_{-\infty}^{\infty} \sigma^2(y) \, p_G(y) \, dy - \mu_{j+1}^2$$

Because of the Gaussian distribution, the properties of the mapping K are determined by the properties of the activation function σ. Several aspects need to be considered in the design of σ: (1) Positive and negative values are needed to adjust the mean. (2) A gradient greater than 1 is needed to increase the variance. (3) Saturation is needed to decrease the variance. (4) Convergence of moments towards fixed points needs a continuous curve.

A suitable activation function is referred to as SELU (scaled exponential linear unit) and reads [35]:

$$\text{SELU}(x) = \lambda \begin{cases} x & \text{if } x > 0 \\ \alpha \, (e^x - 1) & \text{if } x \leq 0 \end{cases} \tag{7.20}$$

The function is similar to the exponential linear unit (ELU) [34], proposed to provide gradients with improved normalization capacities and a robust saturation regime to prevent a randomized on- and off-switching of activations as observed for ReLUs. To this concept, the SELU adds two parameters λ and α. Depending on the initialization of the network parameters, λ and α are adjusted to fulfill the required mapping properties, including convergence towards fixed points. For an initialization based on the normal distribution with mean zero and variance one (respectively $\sigma = 1/\sqrt{n_{\text{in}}}$), the parameters are:

$$\alpha \approx 1.6733 \qquad (7.21)$$

$$\lambda \approx 1.0507 \qquad (7.22)$$

Figure 7.6(a) shows a graph of this activation function. The convergence of the moments to fixed points is ensured by the norm of the Jacobian matrix of K being less than 1. For detailed information and general proofs, the reader is referred to [35]. A visualization of the convergence of the activation distributions is shown in Fig. 7.6(b). Note that neural networks with SELU activation function further need a specialized dropout, so-called *alpha-dropout*, to preserve normalized activations.

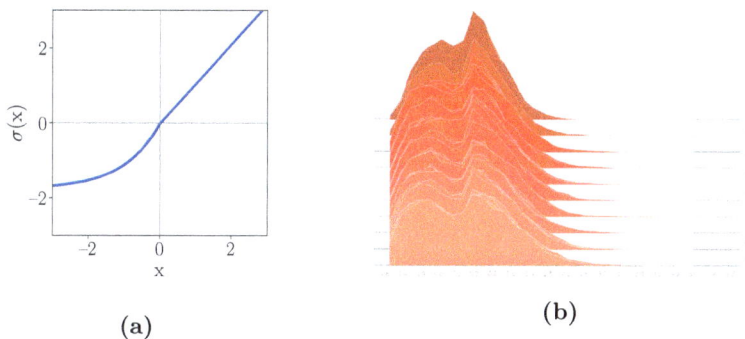

(a)

(b)

Fig. 7.6. (a) SELU activation function preserving zero mean and variance one. (b) Distribution of activations (horizontal) as a function of the training epoch (from back to front).

7.8   Exercises

Exercise 7.1. Classification of magnetic phases: Imagine a two-dimensional lattice arrangement of $n \times n$ magnetic dipole moments (spins) that can be in one of two states ($+1$ or -1, Ising model). With interactions between spins being short ranged, each spin interacts only with its four neighbors. The probability to find a spin in one of the orientations is a function of temperature T according to $p \sim \exp{(-a/T)}, a = \text{const.}$

At extremely low temperatures $T \to 0$, neighboring spins have a very low probability of different orientations, so that a uniform overall state (ferromagnetic state) is adopted, characterized by $+1$ or -1. At very high temperatures $T \to \infty$, a paramagnetic phase with random spin alignment results, yielding 50% of $+1$ and 50% of -1 orientations. Below a critical temperature $0 < T < T_c$, stable ferromagnetic domains emerge, with both orientations being equally probable in the absence of an external magnetic field. The spin-spin correlation function diverges at T_c, whereas the correlation decays for $T > T_c$.

Start with the example script (Exercise 7.1) from the exercise page. The data for this task contain the $n \times n$ dipole orientations on the lattice for different temperatures T. Classify the two magnetic phases (paramagnetic/ferromagnetic) and

(1) evaluate the test accuracy for a fully-connected network,
(2) plot the test accuracy versus temperature.

Summary 7.1.

(1) Fully-connected networks are classic network architectures. By adjusting the inputs and outputs, they can be flexibly adapted to solve very different tasks.

(2) For the prediction of multidimensional continuous values, the last layer holds n output nodes, where each node corresponds to one predicted value. For example, for the spatial localization of events on a surface (two-dimensional vector), $n = 2$ output values are required. As an objective function for network training, the *mean squared error* function is mostly used.

(3) If the network prediction requires an assignment of objects to k categories or *classes*, the number of classes corresponds to the number k of output values. The output values are further processed by the function *softmax* so that each of the k output values can be regarded as a probability that an object belongs to a particular class. *Cross-entropy* is used as the objective function in the network training.

(4) For deep networks, new challenges arise. The training process needs to be protected against the problems of *vanishing or exploding gradients*. Coupling of the gradient of weight W_{ij} with weights and gradients from other layers requires strategies in choosing appropriate learning rates (Sec. 4.8). A problem arises when iterative updates of parameters are performed simultaneously, but in an uncoordinated way, leading to *internal covariate shifts*.

(5) One possibility for stabilizing the training is to optimize the networks by learning *residual* mappings. This can be achieved by modifying the network architecture using so-called *shortcuts*.

(6) A further possibility is re-normalizing a minibatch's training data before entering the next network layer. This is called *batch normalization*.

(7) A third option is *self-normalizing networks*. Here, the distributions of activations in the network layers are iteratively stabilized. Using suitable activations (e.g. SELU), the convergence of the moments (mean, variance) of the activation distributions to fixed points is achieved.

Chapter 8

Convolutional neural networks and analysis of image-like data

Scope 8.1. This chapter presents the concept of convolutional neural networks, which provide a technique to exploit spatial symmetries in data. The key point is to analyze small neighboring segments in the input data to be sensitive to strong local correlations and to utilize translational invariance. We compare the standard convolutional network with the fully-connected network. We also show advanced variants of convolutional architectures and exemplary applications in physics research.

8.1 Convolutional neural networks

For data structures with inherent symmetries, dedicated network architectures exist, which have proven to be far more efficient than fully-connected networks described above.

This chapter focuses on data that have a regular geometric structure like pixels of an image. When discussing properties of image-like data, i.e. the image structure or pixel values, we refer to the term *spatial domain*. This allows a clear distinction from the feature space (Chapter 6). Certain patterns and structures in the image can have a certain meaning. For example, several pixels in a gray round shape form a tire. Two such tires and a wire frame, on the other hand, form a bicycle. The recognition and determination of objects are usually unambiguous for us humans. But such an interpretation is detached from the spatial domain. We refer to the actual meaning of given pixel collections as their *semantic meaning*.

In many experiments, the sensors for physics measurements are placed on a discrete and regular grid, forming domains with spatial symmetries similar to images.

If one processed image-like data using a fully-connected network, the value(s) for each pixel would be input in the neural network. Applying the linear transformation during the transition to a subsequent layer implies that each pixel is correlated with each other pixel of the image. Even for images with moderate resolutions, this results in an enormous number of adaptive weights.

Owing to the spatial symmetry and other regularities in the data, one expects pixels with a similar semantic meaning to populate neighboring areas on the image. For example, many blue pixels in the upper part of an image form the sky.

Similarly, several signals observed in adjacent sensors could belong to a single particle traversing a detector. Here, we would like to answer the question: Was there any particle traversing the detector, or was it noise from the sensors? The question corresponds to a binary classification task of signal against background. In general, for this decision, it is irrelevant where exactly the particle path is located in the detector. A shift of the particle path should leave the classification result unchanged.

Concepts for taking this so-called *translation invariance* into account were developed in the 1990s in the form of Convolutional Neural Networks (CNNs) [36]. Although the basic concept is nearly 30 years old, the breakthrough of CNNs began around 2012 [37] and they are now one of the driving forces in deep learning and computer vision.

In the following section, we will present a phenomenological approach for understanding the CNN concept. A more mathematical definition will be given later in Chapter 10, where we will generalize the concept of convolutions to non-Euclidean domains and graphs.

8.1.1 *Convolution of 2D image-like data*

The basic idea of the CNN architecture is to first analyze images in segments of a few pixels and to subsequently merge the information from all segments. The advantage of this approach becomes apparent when comparing it with the instant analysis of the entire image through a fully-connected network. Here, mapping all pixels to each other requires a very large number of parameters. In contrast, the analysis of image segments can be performed with a much smaller number of parameters.

The localized evaluation of images is realized with the help of *filters* that slide over the image. The weight parameters of the filters are first learned from training data and are fixed during the evaluation of validation

(a)

(b)

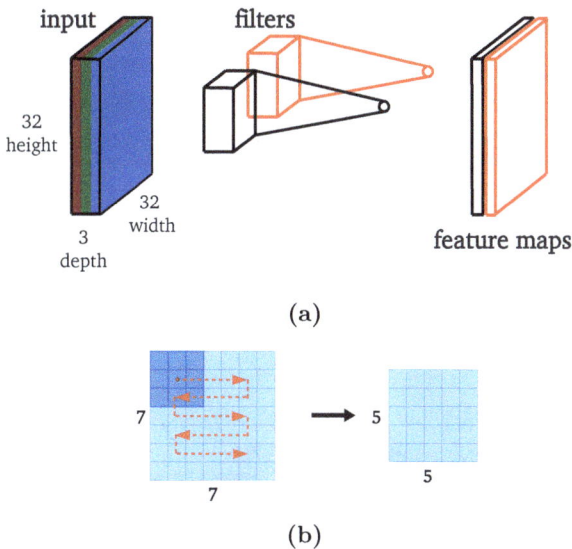

Fig. 8.1. Key concept of a convolutional layer. (a) Two convolutional filters are applied to an input image producing two feature maps. (b) Example movement of a single filter over an input image.

or test data. However, it should be noted that with adaptive filters of small sizes, a prior on the high correlations in the neighborhood of the pixels is realized, as each pixel is only linked with its immediate neighbors.

In Fig. 8.1(a), the concept of convolution using two filters for an image with three color *channels* (RGB=red, green, blue) is illustrated. The filters carry adaptive weights and are shared and evaluated over the whole image. In a figurative sense, each filter moves over the image, as depicted in Fig. 8.1(b). At each filter position, the pixel values are weighted with the respective filter weights. The calculated values are added up over the respective neighborhood, resulting in a single value per position.

Assuming a single-channel image the transformation for a filter at each pixel position i looks similar to the fully-connected network:

$$x_i = \sum_{x_j \in N_i} W_j \cdot x_j + b. \tag{8.1}$$

Here x_j are the N_i pixel values in the neighborhood[1] of pixel i covered by the filter, W_j are adaptive weights, and b is the adaptive bias (compare

---

[1]Including the pixel i itself.

with (3.1)). Note that x_j and W_j are in general vectors as we are dealing with multi-channel inputs. For example, for an RGB input (image with three color channels), the filters can have an arbitrary width and height but have a depth of 3 and are acting on all input channels simultaneously. Furthermore, the single bias b is added to increase the model capacity. Finally, to increase the expressiveness of the model even further, a non-linearity is applied to each pixel response by using an activation function σ (compare with (3.4) and Fig. 5.8):

$$y_i' = \sigma(x_i') \qquad (8.2)$$

An example of calculating the convolutional operation at a single-pixel position is shown in Fig. 8.2. In summary:

> *During a convolutional operation, the information at a given filter position (pixel) is aggregated over the respective pixel neighborhood to a single value by performing a linear transformation, summing over the neighborhood, and applying a nonlinearity.*

Exploiting translational invariance: It is very important to note that the adaptive parameters W_j and b in (8.1) are independent of the filter position (they do not depend on i), as the same parameters are applied all over the image. This concept is called *weight sharing* and ensures the translational invariance of the convolutional operation.

Another important aspect is that using weight sharing, the total number of adaptive parameters does not depend on the number of inputs, as is the case for fully-connected networks. This saving of parameters is a key feature of convolutional networks and is motivated by the translational symmetry of the data.

Feature maps: The result of the convolutional operation using a single filter is referred to as *feature map* or *channel*. This feature map keeps the image-like data structure since sharing the filter over the image, calculating the response, and adding a bias does not affect the pixels' neighborhood relations. As each filter generates its own feature map, one can vary the number of used filters to change the number of feature maps. In this way, one can adapt the amount of information retained for further analysis.

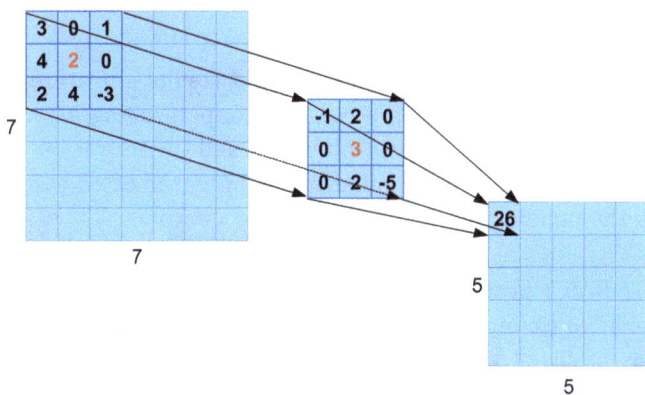

(a)

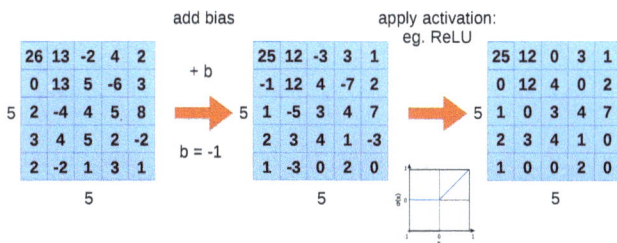

(b)

Fig. 8.2. Example of a convolutional operation.

Example 8.1. Parameter sharing: Imagine that the used filter-size of the convolutional layer would amount to the exact size of the input image. Then each pixel is linked to each pixel in the image and is scaled with an individual weight. Hence no weight sharing would be possible. The resulting feature map of one filter, which is furthermore shifted by a single bias, is then identical to a fully-connected layer with a single node. In general, the number of used filters would amount to the number of nodes in the fully-connected case. This demonstrates that the fully-connected layer can be expressed using a convolutional layer and that moderate filter sizes are essential for exploiting the spatial symmetry of the data.

8.1.2   *Convolutional layers*

Applying a convolution to an image-like input results in feature maps that a subsequent convolution can process. Hence, performing a convolutional operation with n_f filters of given filter size and an activation function σ defines a *convolutional layer*. Similar to the fully-connected case, the layer represents the building block of convolutional networks.

Example 8.2. Parameter counting: Let us further inspect the initial illustrative example in Fig. 8.1(a) where a convolutional layer with two filters of size 3×3 is applied to an RGB image. Hence, for each pixel, the information of its eight neighbor pixels and its own pixel value is collected in the three color channels, weighted by the adaptive filter weights, and aggregated to a single response which is subject to a bias and an activation function σ. The convolution reads:

$$Y'_{i,j,f} = \sigma \left(\sum_{k=-1}^{1} \sum_{l=-1}^{1} \sum_{c=1}^{3} W_{k,l,c,f}\, X_{i+k,\, j+l,\, c} + b_f \right) \qquad (8.3)$$

Here, i and j denote the positions of the input X and output Y pixel, respectively, c the individual color channel, and f is the filter index. The indices k, l describe the aggregation over the neighborhood using the 3×3 filter. As each filter holds $3 \cdot 3 \cdot 3$ adaptive weights, the complete operation to create a single feature map in the convolutional layer only requires $27 + 1 = 28$ adaptive parameters, where the addition of 1 is due to the bias b. Thus the convolutional layer with $f = 2$ filters contains in total 56 parameters in contrast to $(32 \cdot 32 \cdot 3 + 1) \cdot 2 = 6{,}146$ parameters needed for a corresponding fully-connected neural network applied to a 32×32 pixel image.

An important factor in determining the efficiency of CNNs is the number of parameters. Note that, though the filter is shifted over the image pixels and not along the three color channels, each color channel in the filter has its separate weight parameters. Hence, the number of adaptive parameters in a convolutional layer amounts to:

$$n_{\text{params}} = n_c \cdot n_w \cdot n_h \cdot n_f + n_f, \qquad (8.4)$$

where n_c is the number of input channels, n_w the width, n_h the height of the filter, and n_f the total number of applied filters. As each filter creates its own feature map, the resulting number of feature maps amounts to n_f.

An important aspect of (8.4) is that the number of parameters depends only on the number and the size of filters but is independent of the resolution of the image input. See Example 8.2 for a detailed example calculation.

8.1.3   *N-dimensional convolutions*

The concept of convolutions we introduced, to extract patterns out of natural images that feature spatial symmetries, can be extended to N-dimensions. The basic idea remains the same, a small N-dimensional filter scans channels of an N-dimensional regular grid and creates N-dimensional feature maps. Especially in high-dimensional data structures, there may be other important symmetries or correlations between certain dimensions. To allow the model to support the symmetry in data or extract important correlations, it can be advantageous to choose appropriate filter sizes or alternatively share filters over specific dimensions (see Example 8.3). Be aware that currently only a few deep learning frameworks support convolutions beyond three or four dimensions.

Example 8.3. Parameter sharing: Capturing symmetries of the data in the model is of great importance. For example, imagine a regular 2D grid of sensors which measure specific signal traces, i.e. amplitudes $A(t)$ as a function of time t. We assume that the traces show similar patterns, such that sharing the parameters in the spatial dimensions may be advantageous. In this example, the data would have a structure of (x = position in x-grid, y = position in y-grid, $A(t)$ = measured signal traces). An appropriate filter then would be of the shape $(1, 1, n_s)$ where $n_s > 1$ is the respective filter size to analyze $A(t)$. Now, the filter is shared over all sensors to analyze similar patterns, reducing both the risk of overtraining and the number of parameters.

8.1.4   *Fundamental operations in CNNs*

In the following, we introduce a few important concepts relevant for the practical construction of CNNs.

Padding: The attentive reader will have noticed that preserving the image size in the present setup is not possible, as a filter with local extent will never be applied to the edges of an image, which results in decreasing image size [Fig. 8.1(a)]. To prevent the decrease of the image resolution,

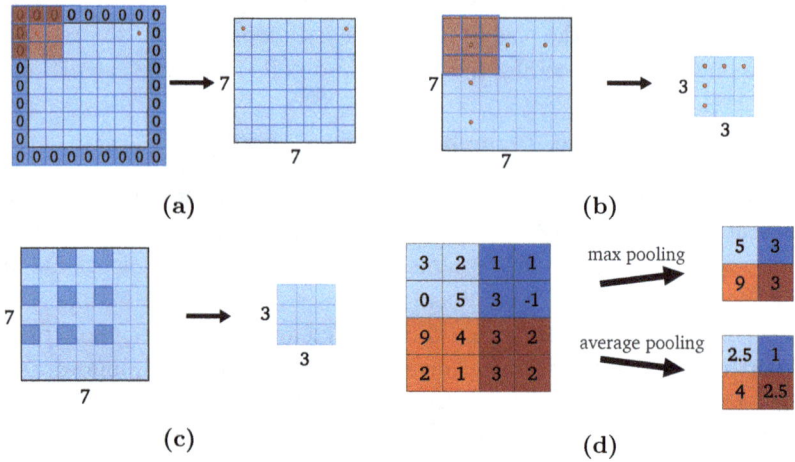

Fig. 8.3. Illustration of basic operations used in CNNs. (a) Padding with zeros is used to conserve the image size. (b) A filter of dimensions 3×3 is acting with a stride of 2, reducing the resolution of the image to 3×3. (c) Dilated convolution with a dilation rate of 1 of a 3×3 filter is used to reduce the image resolution aggressively. (d) Working principle of max- and average pooling to downsample the input image from a resolution of 4×4 to 2×2.

zero padding is used. In the padding operation, illustrated in Fig. 8.3(a), additional zeros are added at the borders of the image, allowing us to place the center of the filter at the image's edges.

Striding and dilating: During the convolutional operation, the feature maps' size can be further controlled by increasing the size of steps between filter applications. In the previously discussed examples, the filter was applied to each pixel. It can be beneficial for very large input sizes or subsequent layers to apply the filter only every n steps. This variant is called *strided convolution* and illustrated in Fig. 8.3(b) for a stride of 2.

Even more aggressively, the image size can be reduced by using *dilated convolutions*. In this particular operation, the filter [Fig. 8.3(c)] has gaps that rapidly reduce the image size. This method is used especially when a large receptive field of view is needed within a few layers. An example application can be a network performing on light-weight hardware. The dilation rate, i.e. the size of holes inside the filter, is another hyperparameter to be adjusted.

Pooling: In contrast to strides and dilated convolutions discussed above that modify the underlying convolutional operation, *pooling* can be used as a form of downsampling. In a patch of size $a \times b$ as defined by the user, a single operation aggregates the information in the neighborhood. Pooling is commonly used with strides of the patch dimensions so that in each pooling layer, each pixel is considered only once, i.e. the patches do not overlap. Commonly used operations are *max pooling* and *average pooling*, which downsample the image by calculating the maximum or mean value in the respective patch, as depicted in Fig. 8.3(d).

The concept of pooling can be extended by defining the patch as the entire feature map. Thus it reduces each feature map to a single value. This technique aggressively reduces the number of outputs and is commonly used at the very end of convolutional architectures. It is referred to as *global pooling*. As entire feature maps are reduced to a single value, in a classification task the operation enforces correspondence between feature maps and categories.

In an illustrative way, pooling can be understood as an operation to merge features of similar semantic meaning (e.g. clustering several blue pixels representing a part of the sky). Further, especially in the case of max pooling, downsampling of images can prevent overtraining, as taking the maximum over a given patch will help the CNN to be less sensitive to particular translations and orientations.

Transition to fully-connected networks: After applying several convolutional and pooling layers, one may assume that a sufficient number of features have been extracted and no further information is encoded in the arrangement of the pixels. Still, the learning of a probabilistic mapping between these features could enhance the capacity of the model. Learning these mappings is similar to adding fully-connected layers to the model, as discussed in Sec. 7.1. For the basic fully-connected layer, the order of the inputs does not matter, such that the multidimensional tensor of feature maps can be reshaped into a vector. This reshape operation is often called *flatten*. It does not change the individual values but only the shape of the objects passed to the next layer.

Example 8.4. Overtraining in CNNs: After several convolutional layers and pooling layers of a convolutional network, fully-connected layers finalize the model. Note that the number of weights for fully-connected layers scale with the number of inputs. If the convolutional layer passes several feature maps with high resolution, succeeding application of the fully-connected layer will produce an enormous number of parameters that may be prone to overfitting. Thus, for an appropriate reduction of the image size, regularization, or a global-pooling operation can help to overcome this problem. The global-pooling method can be viewed as a fully-connected transformation with a diagonal weight matrix.

8.1.5 *Learning short- and long-range correlations*

In a convolutional network, complex operations are divided into many low-complexity operations (layers), forming the foundation of so-called *hierarchy learning*. We have already experienced this with fully-connected networks.

In each convolutional layer, only the immediate neighbors are linked. Thus one could argue that with a CNN, long-range correlations may not be represented. However, in each convolutional layer, every single response in a feature map consists of aggregated local neighborhood information. In the subsequent layer, where we apply a convolutional operation to the feature map responses, the direct neighborhood of the applied filter is considered, and all neighborhoods that have been aggregated in responses from preceding network layers. Thus, in each subsequent layer, the region where relationships can be exploited is continuously increasing. In the terminology of computer vision, this is referred to as the *receptive field of view* and specifies the part of the input image that is visible to a filter at a specific layer [Fig. 8.4(a)].

The increasing receptive field of view with an increasing number of layers can be seen as a perfect example of *hierarchy learning*. Learning of short-range correlations, lead to more semantically-global features appearing in the first layers. Long-range correlations and, as a consequence, more detailed and (task-) specific features are learned in later layers. The hierarchical understanding of CNNs will be detailed in Chapter 12. Hierarchy learning is supported by a basic convolutional architecture proposed in AlexNet [37], which has been a milestone in applying CNNs in

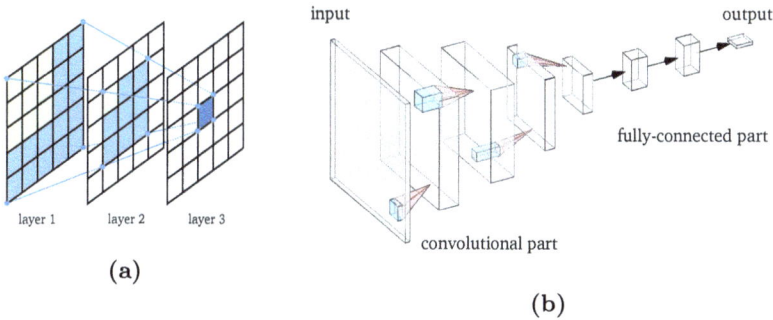

Fig. 8.4. (a) Increase of the receptive field of view for deeper layers. (b) Convolutional 'pyramid' defined by a decrease of spatial dimensions and an increase of feature dimensions [19].

computer vision. In this architecture, whenever the spatial dimension is reduced, the number of features is increased correspondingly, which leads to a pyramid-like structure visible in Fig. 8.4(b). In a figurative sense, the concept extracts *features* encoded in the *spatial* structure of the image and stores them in *feature space*. This reduces the computing effort while still achieving good results.

8.2 Comparison to fully-connected networks

To enhance our understanding of the convolutional operation, we compare the weight matrix of a convolutional layer with a single filter to that of a fully-connected layer. In Fig. 8.5, the weight matrix corresponding to a 1D convolution is shown. We assume the input data exhibit a regular structure such as a measured signal trace of an amplitude A in time intervals Δt of a fixed size.

In the example of Fig. 8.5, the input trace has five time steps. The filter has a length of 3. As we can apply this filter five times (operation with padding), the corresponding weight matrix $\mathbf{W}$ for the linear mapping of a fully-connected network layer has the shape 5×7. As only neighboring pixels are considered in the 1D convolutional operation, the convolutional layer's weight matrix has a diagonal-like structure. In the outer part, the matrix does not contain adaptive weights. The weights are further reduced as the same filter is shared over the signal trace. Thus, in each row, the same parameters are used, which reduces the number of parameters

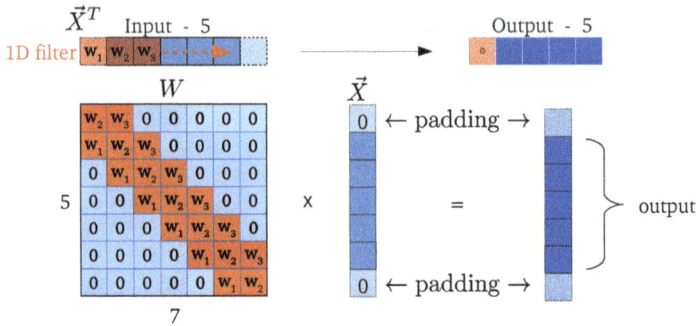

Fig. 8.5. Matrix representation of a 1D convolution. The number of parameters is drastically reduced due to symmetry considerations without affecting the degree of model complexity.

massively compared to fully-connected networks where the entire matrix holds adaptive weights.

The special feature of the convolutional operation is that the reduction of weight parameters arises from symmetry considerations. Therefore, the descriptive power of the model is not reduced when applying CNNs to data that exhibit such a regular structure. The resulting major advantage is that by reducing the number of parameters, the related numerical optimization problem is greatly simplified. Furthermore, the risk of overtraining is reduced since weight sharing stabilizes the learning of filter weights, as single filters are much less susceptible to fluctuations and artifacts.

Example 8.5. Parameter sharing: To understand the power of weight sharing and the learning of filter weights, let us assume that we want to detect a cat. Imagine during the training of the CNN one filter was learned, which produces a high response when encountering the face of a cat. Whenever this response is high, we can identify a cat. As in our convolutional model, the filter is shared over the image. It does not matter where the cat is in the picture. Wherever the filter matches the face of the cat, it will produce the same large response. This allows us to reduce the number of parameters efficiently by using the symmetry of our data. It can even help to find the cat's position in the image by looking where on the feature map of our filter the highest activation appears (see discriminative localization, Sec. 12.3.2.1).

8.3   Reconstruction tasks

The underlying structure of image-like data allows the study of various reconstruction tasks. For instance, the following questions may be addressed:

- Classification: Is the detector response signal or background?
- Regression: What is the energy of a measured particle?

These tasks can also be performed at the pixel level. Here, certain values or classes are assigned to individual pixels or parts of the input. Examples for additional tasks are *pixel-wise regression*, *object localization*, and *semantic segmentation*. While in object localization, one tries to only roughly locate the position of individual objects, this is done in semantic segmentation at the pixel level.

Semantic segmentation: Figuratively, the task of semantic segmentation can be seen as *pixel-wise classification*. This means that the network outputs do not need to consist any longer of output nodes but output maps with the number of possible classes. Each map represents one class and usually has the resolution of the input image. Each respective pixel value in a particular map states the probability of the related input pixel belonging to the respective class. By finding for each pixel the most probable class results in a pixel-wise segmentation of the input image.

Pixel-wise regression: While classification is more common in the literature, regression for image-like data has so far played only a subordinate role. In principle, it is also possible to perform regressions on the pixel level. Emerging fields of physics applications are featured in Examples 8.6 and 8.11.

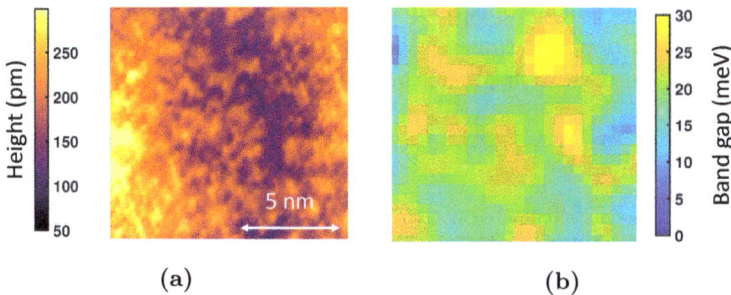

Fig. 8.6.   Scanning probe microscopy of the same area of a $MnSb_2Te_4$ epitaxial film: (a) topographic image, (b) local band gap [38].

Example 8.6. Scientific imaging: Imaging is one of the most important scientific tools to extract local information in the spatial domain. In today's research, special emphasis is placed on gaining deeper insights through the correlation of images that show different properties, and on the increase of resolution. A prime example of the first aspect are scanning tunneling microscopy techniques, which allow to scan the same area with different bias voltages. Depending on the height of the bias voltage, the resulting image can exhibit the topography of the scanned area, i.e. an essentially geometric property, or local electronic properties of the same area [39].

For example, Fig. 8.6(a) shows an atomic resolution topograph of a $MnSb_2Te_4$ semiconductor layer of 100 nm thickness, deposited on a BaF_2 substrate. The image was taken at a fixed bias voltage of $V = 500$ mV and a tunneling current of $I = 120$ pA. The long-range corrugation observed in the image is caused by an inhomogeneous distribution of Mn atoms below the surface [38]. At each point of the scanned area, local electronic properties can be determined from the variation of the tunnel current in response to a change in bias voltage. In particular, the local differential conductivity $dI/dV(\vec{r}, V)$ is a measure of the density of electronic states at the tip position. Measured for a suitable range of voltages near the Fermi level, the dI/dV spectrum can indicate specifically the local size of the band gap, characterized by a lack of available states. This can be used to generate an image that displays the local band gap [Fig. 8.6(b)]. The identification of correlations between such images is highly desirable to deepen the theoretical understanding of the relationship between structure and properties, and would also be of great practical benefit. For instance, the pixel-wise prediction of the local density of states from a topograph would allow accelerating experiments, reducing the need for time-consuming scans at different bias voltages. Moreover, samples with interesting, unusual properties could be identified faster.

8.4 Advanced concepts

In recent years, many advanced concepts have been developed in the context of convolutional networks. Below, we present several improvements which are particularly interesting for applications in physics.

Point-wise convolution: The point-wise convolution is a special case of the convolutional operation. In contrast to the standard convolution, which focuses on local correlations in the spatial domain, point-wise convolutions act only in the feature domain. Using filter sizes of (1×1), each feature map is scaled by a single weight value. Thus, the transformation can be seen as a fully-connected layer applied to the features space.

During the transformation, usually the number of feature maps is reduced. The purpose is to merge a large number of feature maps to fewer but more abstract ones. Since they are used to produce low-dimensional representations of the data and reduce the computing effort, they are often called *bottleneck layers*.

Locally-connected convolutions: *Locally-connected convolutions* are similar to the normal convolution operation, but the filter weights are not shared over the image. Instead, each local patch of the input is processed by a different filter. This operation is therefore not invariant under translations. It can be useful if the data show strong local correlations, but the symmetry of translational invariance is broken.

Batch normalization: CNNs are relatively robust against noise since they share the kernels over a complete image. Therefore, they are ideally suited for extensive use of batch normalization (see Sec. 7.6), together with ReLU activations.

Inception network: Within a convolutional layer, the size of the filter determines the size of the structures that can be learned. Inception modules [40] are used to increase the sensitivity within a network layer to patterns of different scales. Here several convolutions with different filter sizes operate in parallel. Note that after every single convolution, an activation function is applied.

After branching, the learned feature maps are combined again via concatenation and form a common feature space that subsequent modules can process. This concept prevents the learning process from adapting to similar features in the various filters.

In Fig. 8.7, an exemplary inception module is shown, which uses four different filter sizes. To prevent the number of feature maps from becoming too large, multiple bottleneck layers (point-wise convolutions) are used. For simplicity, the activation is not shown but is usually applied after each convolution.

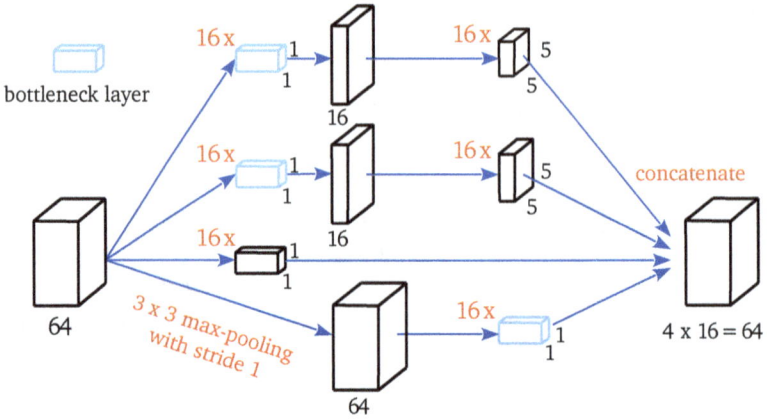

Fig. 8.7. Working principle of an inception module. A single convolutional operation is broken down into a couple of convolutions working on different scales.

Separable convolutions: In a figurative sense, we motivated the Inception network with features on different scales. Another perspective of the Inception's success is that if *spatial correlations* and *cross-channel correlations* are sufficiently decoupled, it is more efficient to calculate them separately. Recently published works use this decoupling further to make convolutional operations more efficient and increase the capacity by enabling to learn powerful channel-correlations [41,42].

Separable convolutions can be seen as an extreme variant of Inception. A variant, referred to as Xception, is illustrated in Fig. 8.8. The basic idea is to decompose the standard convolutional operation by factorizing it into two subsequent convolutions. The first only acts on the spatial dimension of each respective channel individually, which is also known as *depth-wise convolution*, before cross-correlating the channels using a second, point-wise (1×1), convolution.

Overall, the technique shows two distinct differences to Inception modules. First, the absence of a nonlinearity between the channel-wise convolution and the point-wise convolution. Experiments show that adding an activation here even reduces the performance. Secondly, the order of the operations is changed, but it can be assumed not to have a large influence due to the immediate following matrix multiplication one after the other.

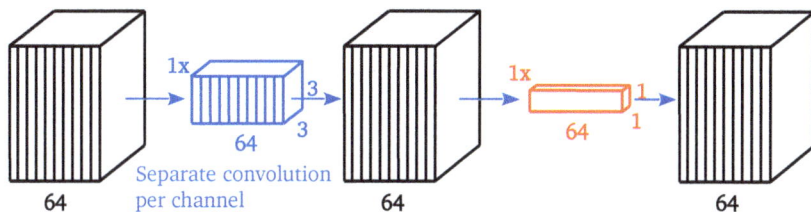

Fig. 8.8. Xception modules consisting of a channel-wise and a point-wise convolution.

Transposed convolutions: In some cases, it is necessary to reverse the effect of a regular convolution.[2] For this purpose *transposed convolutions*, sometimes called *deconvolutions*,[3] are used.

In the transposed convolution we apply a filter that acts on a single pixel on the input image but produces a patch as the output. By sliding over the input, several of such (overlapping) segments are created. Finally, by aggregating over the produced patches the output is obtained (see Fig. 8.9).

Speaking simply, this is the contrary operation to the regular convolution and one can think of going from caused activations back to the input image. This change of operation creates some effects that may seem strange at first. Whereas in the standard convolution, strides > 1 reduce the resolution in the output, in the transposed case the resolution will be increased. To explain this effect, consider the strides to be applied to the output and not the input, or, to insert gaps into the input image. Also, padding is now rather defined on the output than the input. So, a transposed convolution *without* padding translates to zero padding in the input (see Example 8.7).

A wide range of applications of this technique can be found in semantic segmentation, autoencoders, and generative models.

---

[2]This is actually always the case when training CNNs, as the operation changes its direction during the backward pass.

[3]Note that the word 'deconvolution' is somewhat misleading. The inversion of DNNs is the subject of current research and usually limited to specific network designs (e.g. see Fig. 18.10) or approximations, as, e.g. the inversion of ReLUs or pooling operations lacks uniqueness.

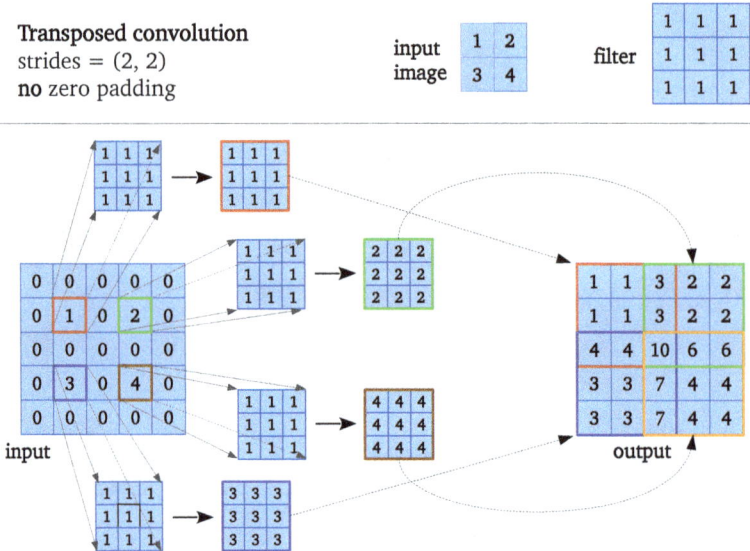

Fig. 8.9. Example of a transposed convolution applied to a (2×2) input with a (3×3) filter, no zero padding and a stride of 2. The application of the filter yields four patches, one for each image pixel, which are aggregated to an output with increased dimensions (5×5).

Example 8.7. Transposed convolutions: In Fig. 8.9, we present a transposed convolution using a 3×3 filter on a 2×2 input. Here we use zero padding and a stride of 2, which yields four patches, one for each image pixel from the product of the filter with each image pixel. This filter application differs from the regular convolution, as one pixel produces one patch. The summation of the four overlapping patches gives the final result.

To understand the definition of strides and padding in transposed convolutions, one can interpret it for the above example (strides = 2, and no padding) in the following way: A regular convolution to a 5×5 image with strides of 2 and no padding would result in a 2×2 feature map.

Upsampling: Another method to change the image size is the so-called *upsampling* operation. Here, input values are simply repeated in a patch with variable size to be defined. Thus, it can be understood as the opposite operation to pooling. In many applications instead of transposed convolutions, upsampling is used followed by a convolution, as sometimes transposed convolutions tend to create checkerboard-artifacts. Again, a wide range of applications can be found in the field of semantic segmentation, autoencoders, and generative models (Chapters 17 and 18).

8.4.1   *Shortcuts*

Residual networks: The basic idea of the residual network (ResNet) [43] is to learn only small changes in each network layer. In Fig. 8.10(a), the concept of residual modules is visualized. The module divides the operation into two parts: the pass-on of the original input $\mathbf{x}$ and a small change $f(\mathbf{x})$ resulting from a few convolutional layers. The output of the network operation is an addition to the input $\mathbf{x}$ and the network result $f(\mathbf{x})$ (compare with Sec. 7.5):

$$\mathbf{y} = \mathbf{x} + f(\mathbf{x}) \tag{8.5}$$

The effectiveness of ResNet can be illustrated as follows. Using many small mappings, a very large capacity can be created in the network. The input $\mathbf{x}$ has priority in every network layer because already adequately learned features are easily propagated throughout the network by zeroing the weights of the branch $f(\mathbf{x})$. Even when using many network layers, the

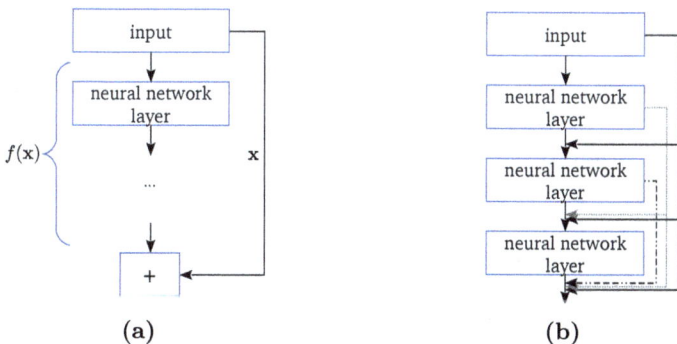

(a)   (b)

Fig. 8.10. Shortcuts as used in (a) residual networks and in (b) densely-connected networks.

network training is less disturbed by deeper layers compared to a standard CNN. The latter would have to learn diagonal matrices for the weight parameters, which is not the case with ResNet. The bypass also reduces the potential problem of vanishing gradients.

With ResNet's massive use of shortcuts between layers of different hierarchies, network architectures can be created that enable training very deep networks with more than a thousand layers and offer an extremely high capacity for complex problems.

Densely-connected convolutions: A related method is densely-connected convolutions used in DenseNet [44]. The structure of these modules is visualized in Fig. 8.10(b). In this technique, shortcuts are also used, but the decisive difference is that the operation of addition is replaced by *concatenation*.

A concatenation operation of two feature maps of dimension $m \times n$ results in a new feature map of twice the dimension, i.e. $2 \times m \times n$. In densely-connected convolutions, all feature maps resulting from preceding network layers are linked together after each layer. In this way, different levels of hierarchy are related to each other.

This combination can also be understood as a kind of feature sharing (similar to weight sharing) that generally stabilizes the training of such networks. Besides, it allows a simplified propagation of gradients.

However, merging feature maps from different levels leads to a rapid increase in memory consumption. To reduce this, different DenseNet modules with multiple layers are often interlinked, using point-wise convolutions as transition layers in between.

UNet architecture: Common tasks in computer vision are the global classification of images, the recognition of individual objects in images, and the division of images into separate image segments (Sec. 8.3). In science, these tasks are typical challenges in the biological or medical field, where imaging techniques are frequently used.

To distinguish different object locations on X-ray images, a network needs to evaluate the images pixel by pixel, and its output must correspond to the dimension of the image. This requires an enormous computational effort. Beyond this, objects such as bones and cell types have to be classified on the image segments, for which convolutional operations are very well suited. For this task, however, they require comparatively few weight parameters.

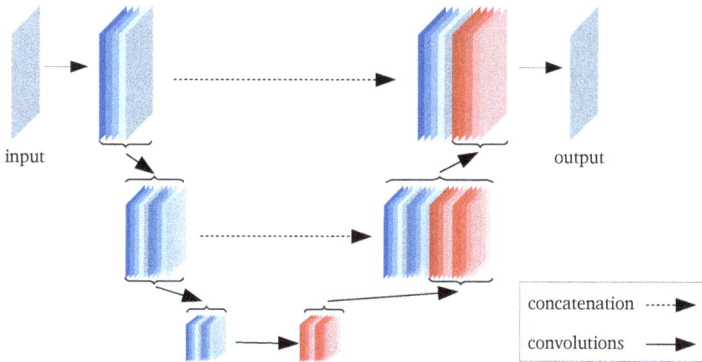

Fig. 8.11. UNet architecture that permits precise object localization and object classification using concatenation operations.

The so-called *UNet* architecture [45] combines these opposite size requirements to enable both precise object localization and object classification. Figure 8.11 shows the two-component UNet design, which consists of a contracting and an expanding part. The first part is quite similar to a standard CNN. After a couple of convolutional network layers and a pooling operation, a dimensionally reduced intermediate representation of the image is achieved. With this information, object recognition could be performed.

The second, expanding part is used to localize objects. Here *transposed convolutions* are used to expand the previously contracted spatial dimensions. It is crucial that after each of these transposed convolutions (or upsampling), the feature maps are combined with the feature maps of the same resolution of the contracting part (*concatenation operation*). Thus, in each network layer of the expanding part, both the compressed feature maps of the first part and the expanded information are available to reconstruct the relevant aspects of the image. A further application of UNet can be found in the field of autoencoders (see Chapter 17).

8.5 Convolutions beyond translational invariance

The successful concept of convolutions, based on strong local correlations and translational invariance, can be transferred to other symmetries. Thus, after considering translational invariance, it is natural to investigate the rotational invariance.

An example of such an application is an architecture named 'Hexa-Conv' [46]. Here the convolution is defined on so-called $p6$ groups, i.e. on hexagonal grids. Besides the weight sharing in the translation, the filter weights are also shared in the rotation. A single channel input is processed by one filter but produces six feature maps, one for each orientation of the filter.

To work with hexagonal data structures, first of all, a coordinate system is required for the localization of the elements (e.g. camera sensors). The so-called axial coordinates, or alternatively the offset coordinates, can be used for this purpose. As shown in Fig. 8.12, the two coordinate systems differ in the position of the second index (check for the arrows). For convolutional operations on the input data, axial coordinates are often preferred. Although they require more memory, filters can be moved more easily across the image.

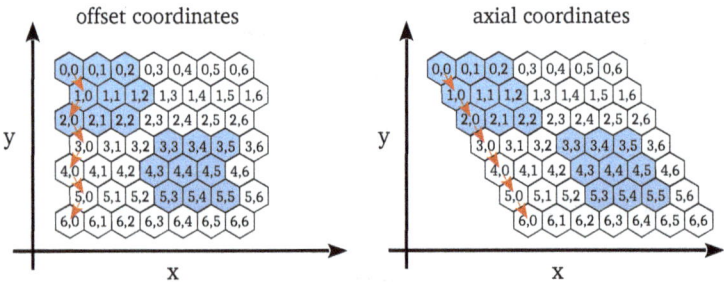

Fig. 8.12. Axial and offset coordinates for hexagonal grids.

In Fig. 8.13, an exemplary application with two convolutional operations is shown. In the first, a single filter is applied to the input data for which only one pixel was activated. Since the filter is rotated, the operation yields six *orientation channels*. Note that these channels are in general not rotated copies of each other. In the second layer, again a single filter is applied. The rotation of this filter and application to the six channels results in six new orientation channels as output of the second layer. This corresponds to a mapping $p6 \rightarrow p6$.

In many experiments, detectors with hexagonal sensor placements are used. It can be advantageous to use this symmetry explicitly in the model, as weight sharing reduces the number of adaptive parameters, which is usually reflected in improved performance.

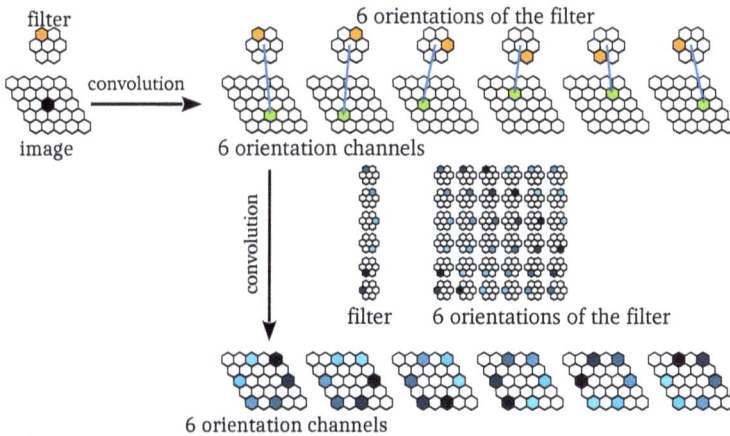

Fig. 8.13. Working principle of the hexagonal convolution visualized using two subsequent convolutional operations. The initial convolution, using a single filter, results in six orientation channels. The following convolution using again a single filter acting on the channels yields another six orientation channels. Due to the simplicity of this example, these channels are just rotated copies of each other.

8.6 Physics applications

The concept of convolutional neural networks was explored early on in physics experiments. Initially, existing computer vision concepts were applied to the situation of the experiments. In current developments, the requirements of the experiments are often in the foreground so that network concepts are developed and adapted accordingly [2]. We refer here to early examples from particle experiments, which provide useful information for understanding how to apply CNNs in physics. A few applications can be found in the following examples.

In particle physics experiments, it is common to generate graphical displays of collision events on the computer (Fig. 8.14). Such event displays show the tracks of particles crossing the measurement sensors of the detectors. An important question here is to determine which particle is observed: electron, muon, ...? An identification of the particles corresponds to a classification, where the classes match the particle types. Often the trajectory in the detector can be used as a criterion for the classification. Sometimes also the signals in the measurement sensors are slightly different for the different particles.

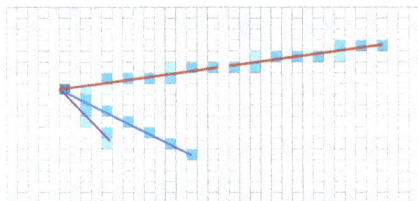

Fig. 8.14. Event display of a particle interaction within a detector.

Example 8.8. Particle identification: Event displays are a natural way to represent experimental data in the form of images that can be directly processed and classified using CNNs. This approach was followed in an experiment on neutrino oscillations [47]. Approximately 800 km from the neutrino generation site, the type of neutrino in question was determined by detecting an electron or muon from such images. While muons leave a ~ 10 m long narrow trace of signals, electron traces are less long and broader.

Using the Inception concept (Sec. 8.4), convolutional filters were trained on the track patterns of electrons and muons. The network improved the identification of electrons by about ~ 30% over previous algorithms.

Example 8.9. Quarks and gluons: At the LHC collider of CERN, particle collisions produce focused clusters of particles (jets) that hit the detector. If the jet originates from an up or down quark, its width is smaller, and the number of particles in the jet is lower than for a jet generated by a gluon. To distinguish between quark- and gluon-initiated jets, the measured particles were projected onto special cylindrical coordinates (η, φ) [Fig. 2.1(c)], resulting in an image-like grid for analysis by a convolutional neural network.

The central idea was to use physics-inspired quantities instead of the RGB channels of a natural image [48]. For the red color channel, the transverse momenta of the charged particles were inserted. For the green color channel, the transverse momenta of the neutral particles were taken. The number of charged particles was fed into the blue color channel. Based on simulation data, the convolutional neural network resulted in a better separation of quark- and gluon-initiated jets than previous algorithms.

Example 8.10. Pixel-wise segmentation of particle trajectories: For the measurement of neutrino particle reactions, huge detector volumes are needed. An efficient technique for > 50 m^3 active volume is the so-called LArTPC detectors (liquid argon time projection chambers). Along the traces of the reaction particles, electrons are released by ionization, which are transported by electric fields to the periphery of the detector volume where they are detected. By time measurements, the particle reaction several meters away, occupying only a few cubic centimeters, can be reconstructed in a three-dimensional (3D) grid [49].

For neutrino measurements in a LArTPC, the UNet architecture (Sec. 8.4.1) was employed to make a pixel-wise segmentation of particle tracks (semantic segmentation, Sec. 8.3) by using measured 2D projections of the underlying physics event. A distinction was made between a track pixel, a shower pixel, and background noise. The error rate in the individual pixel classification was found to be at a low-percentage level.

Example 8.11. Fluorescence microscopy: The resolution limit in conventional optical microscopy is the well-known diffraction limit, a physical barrier that restricts the resolution of image details to about half the optical wavelength. Nevertheless, super-resolution methods have been developed [50], e.g. in fluorescence microscopy. This technique is extensively used in biophysics, allowing the visualization of in-vivo samples in great detail. The techniques employ fluorescent molecules as stochastically blinking emitters for structural analysis at the subcellular level. Since photoactivable molecules are utilized, it is possible to activate only a few molecules at a time, which in turn allows to estimate their positions by a localization algorithm. This so-called *single-molecule localization microscopy* (SMLM) permits the generation of a high-resolution image by integrating many localization outcomes into a single image. Convolutional neural networks have been successfully applied to the localization of fluorescent molecules in 3D, using voxels instead of pixels. In Ref. [51] deep networks were used to decompose the localization problem into a classification task (is there a molecule in a given voxel?) and a regression analysis (what are the coordinates of the molecule inside the voxel?). When applied to real and simulated data, the approach proved to be competitive with standard SMLM methods in terms of computational efficiency while providing a voxel resolution of $24 \times 24 \times 50$ nm^3.

8.7 Exercises

Exercise 8.1. Classification of magnetic phases: We work again with the Ising model of a two-dimensional lattice arrangement of $n \times n$ spins (see Exercise 7.1).

(1) Classify the magnetic phases and evaluate the test accuracy for a convolutional neural network with a global average pooling layer after the last convolutional layer.
(2) Plot the test accuracy versus temperature.
(3) Compare the result to the fully-connected model of Exercise 7.1.

Summary 8.1.

(1) Convolutional Neural Networks (CNNs) are the standard way for building deep networks for image-like data.
(2) CNNs simplify the underlying numerical problem by using symmetries inherent to images.
(3) By sliding small filters with adaptive weights over the input, the convolutional operation can deal with variable input and output sizes and considers:

   (a) Local neighborhoods that are semantically similar: the filter only covers a subpart of the image,
   (b) Translational invariance: the filter weights are shared over the image.

(4) Exploiting symmetry in data allows to reduce the total number of model parameters, i.e. the number of weights is independent of the input resolution.
(5) By stacking convolutional layers, the receptive field of view increases, which allows to extract features of different hierarchies.
(6) More advanced architectures deal with shortcuts and filters with sensitivity to different scales, which can improve the performance and allow to build very deep models.
(7) Using batch normalization in CNNs can increase the models' convergence and performance, especially when training very deep models.

Chapter 9

Recurrent neural networks: time series and variable input

Scope 9.1. This chapter introduces recurrent neural networks as elementary networks for the processing of sequential data such as time series, text documents, or any other form of ordered data. The focus is on general principles and two types of realizations: long short-term memory and the gated recurrent unit.

9.1 Sequential relations and network architecture

Sequential data are inherently ordered. They are encountered in time series, words, phrases, and many other instances of discrete sequences such as the string of nucleobases that make up DNA in genetics. Especially time series are ubiquitous in a world interconnected by large data streams, be it weather forecasts, financial markets data, or all forms of audio and video recordings. Extensive data streams also exist in physics, for example at CERN, where collider experiments produce about 90 petabytes of data per year. Since time evolution is a central theme in all areas of physics, the following discussion of sequential data will be based on temporal sequences. However, the generalization to other forms of ordered data is straightforward.

We start by noting that no matter how well a feedforward neural network performs, it is completely ignorant of the order in which data are presented to its input. Hence, this type of network is not suited to problems which require the use of context, for example making predictions about the next step of a sequence. Even if we create context by concatenating sequential data to a single input vector, the problem remains that in a traditional feedforward network the input is still processed simultaneously in the first hidden layer. When dealing with sequences, it is therefore natural to seek an alternative network architecture that takes into account already stored elements when processing a new one.

This type of behavior is realized by a recurrent neural network (RNN), which is an artificial neural network distinguished by an additional memory called *internal state*. It provides a great deal of flexibility that is otherwise difficult to achieve, such as allowing input sequences of variable length. Although RNNs were already invented in the 1980s [52], computational issues like the vanishing gradient problem (see Sec. 7.4) hindered their practical use until the late 1990s. Nevertheless, their breakthrough was only in the last decade when large data sets became available for training.

9.2   Recurrent neural networks

A basic unit of a recurrent network is shown schematically in Fig. 9.1. Let us assume we measure d different quantities simultaneously in an experiment. Then we have actually d distinct sensor readings at each tick t of the time series, and our input $\vec{x}_t$ is a d-dimensional vector. For simplicity, the d nodes of the input have been replaced by a single one in Fig. 9.1. Note that the processing of the input is represented here as vertical flow, which will allow us later to use the horizontal axis for different time steps. The hidden state at each time t is represented by a p-dimensional vector $\vec{h}_t$. It is computed from both the current input vector and the previous state of the hidden vector at time step $t - 1$:

$$\vec{h}_t = f(\vec{x}_t, \vec{h}_{t-1}) \tag{9.1}$$

The output vector $\vec{y}_t$ could be for instance a forecast of $\vec{x}_{t+1}$. It is an

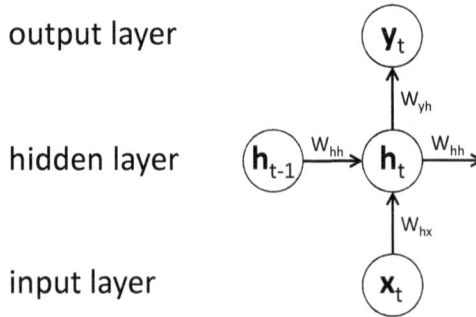

Fig. 9.1. Basic unit of a recurrent network.

n-dimensional vector calculated only from the hidden layer:

$$\vec{y}_t = g(\vec{h}_t) \tag{9.2}$$

The functions f and g remain the same for each time step, thus implying a time series with underlying characteristics that do not change over time.

The computations are performed in the same way as for feedforward networks, e.g. we apply some matrix multiplication with weights $\mathbf{W}$, add some bias $\vec{b}$ and pass the result through a nonlinear activation function. If we use tanh for f and another activation function σ for g (for example, a sigmoid), we obtain for instance

$$\vec{h}_t = \tanh\left(\mathbf{W}_{\mathrm{hx}}\,\vec{x}_t + \mathbf{W}_{\mathrm{hh}}\,\vec{h}_{t-1} + \vec{b}_{\mathrm{h}}\right), \tag{9.3}$$

$$\vec{y}_t = \sigma(\mathbf{W}_{\mathrm{yh}}\,\vec{h}_t + \vec{b}_y). \tag{9.4}$$

We note that the matrix $\mathbf{W}_{\mathrm{hx}}$ is of dimension $p \times d$, whereas $\mathbf{W}_{\mathrm{hh}}$ is of dimension $p \times p$ and $\mathbf{W}_{\mathrm{yh}}$ of dimension $n \times p$. In the above notation, it is understood that the activation function acts element-wise on a vector.

The definition of $\vec{h}_t$ indicates that the hidden vector is updated every time a new data point is fed to the input layer. The recursive nature of (9.1) implies furthermore that the input $\vec{x}_t$ not only interacts with the hidden state from the previous step, but indirectly with input from *all* previous time steps. This results in a function u that depends on all prior input vectors of the time series:

$$\vec{h}_t = f(\vec{x}_t,\, \vec{h}_{t-1}) \tag{9.5}$$

$$= f(\vec{x}_t,\, f(\vec{x}_{t-1},\, \vec{h}_{t-2})) \tag{9.6}$$

$$= u(\vec{x}_t,\, \vec{x}_{t-1},\, \vec{x}_{t-2}, \ldots) \tag{9.7}$$

It is these recursive characteristics that enable RNNs to process input data of variable length.

Figure 9.2 shows a reduced version of Fig. 9.1 restricted to the input layer and the hidden layer. This can be regarded the core building block of an RNN, since an output is not necessarily required at each time step (see Sec. 9.6).

The unit labeled 'A' performs a computation according to (9.3). Its recursive properties are usually symbolized by a self-loop depicted on the right, since information on the hidden state is passed on as additional input from step t to step $t+1$. For a series of finite length, the loop can be unrolled into a layered network as shown in Fig. 9.3, allowing to keep track of each time step. For example, at $t = 0$ the input $\vec{x}_0$ is fed into the network to produce $\vec{h}_0$. During the next time step, input for unit A is not only $\vec{x}_1$,

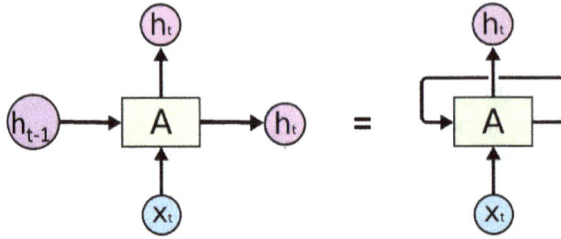

Fig. 9.2. Core building block of a recurrent network and its self-loop shorthand. Note that compared to Fig. 9.1 only the input and hidden layers are shown. Adapted from [53].

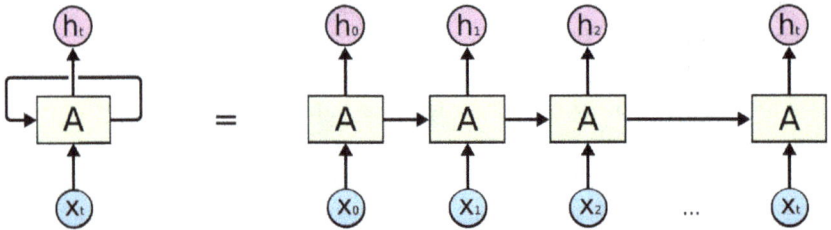

Fig. 9.3. Unrolling of a recurrent neural network [53]. The horizontal axis is equivalent to a timeline.

but also $\vec{h}_0$ from the previous time step, and so forth. The time evolution looks much more familiar when the RNN is represented by such a directed graph. However, the actual similarity to traditional feedforward networks is limited, since the data flow in Fig. 9.3 is horizontal *and* vertical, and the input is received by *all* unrolled layers.

The recurrent unit can be considered an internal memory that is directly controlled by the network. Hence, such stored states are called gated memory, which find use in more complex structures discussed in Secs. 9.4 and 9.5.

RNNs are known to be Turing-complete, i.e. any Turing machine can be simulated by an RNN of finite size [54]. This theoretical property essentially asserts the ability to compute any function if sufficient resources and data are provided. However, its practical benefit is so far quite limited, since long sequences tend to run into generalization problems.

9.3 Training of recurrent networks

Similar to the networks discussed in preceding chapters, RNNs also need training. This can be difficult for long sequences, as these lead to very deep networks. The usual weight update by backpropagation, explained in Sec. 4.5, has to be replaced by a backpropagation process adapted to sequential data, called *backpropagation through time* (BPTT). The BPTT algorithm is based on the unrolled version of the recurrent network and takes into account that all layers share the same parameters, which is evident from Fig. 9.3. In regular backpropagation, a gradient is calculated for each input, which is then propagated backwards through the network from the output to the input to update weights. In BPTT, the gradients for the network parameters are accumulated over all time steps before passing the result backwards through the network.

It should be emphasized that BPTT is prone to the vanishing gradient problem. Although, in theory, RNNs can process any long-term dependencies in their input sequences, severe computational challenges arise in practice during training since the backpropagating gradients approach zero through continued multiplication of small numbers. In essence, the vanishing gradient problem in BPTT is a numerical instability related to the N-fold multiplication of the weight matrix $\mathbf{W}_{hh}$ with itself when dealing with a time sequence of N steps. This inevitably leads to poor long-term memory when N becomes large, no matter how good the short-term memory is. Unfortunately, critical events of time series often turn out to be separated by large gaps, thus escaping the learning process of basic RNNs. Clearly, an extended short-term memory is needed that drastically limits the sensitivity to gap length in sequence learning. Specialized versions of RNNs have been proposed to address this problem. We present two key architectures in the following sections.

9.4 Long short-term memory

The long short-term memory (LSTM) was developed to cope with the issue of long-term dependencies [55]. In particular, it solves the vanishing gradient problem, which is particularly severe in RNNs as outlined in Sec. 9.3. Since its introduction, it has dramatically improved the performance of recurrent networks, and it is fair to say that LSTM networks have paved the way for the practical success of RNNs today.

Unlike standard recurrent networks, LSTM units (referred to as cells) are able to remember data over arbitrarily long time intervals. This is achieved by

- an internal memory called cell state,
- gates that control the cell state actively.

The idea is to modify the recursive formula (9.1) for the hidden vector $\vec{h}_t$, thus allowing better control over its updates. To this purpose, a *cell state* vector $\vec{C}_t$ is established, also simply called cell vector. It is a p-dimensional vector like $\vec{h}_t$, but separate from it. Being the key element of the memory, it can selectively preserve or delete stored elements, and take in new information as well. To that purpose, a common LSTM architecture is composed of a cell and three gates that control the flow of information inside and out of the cell:

- The *forget gate* decides whether to forget or keep elements stored in the cell. Therefore, it is also called keep gate.
- The *input gate* decides whether a new value flows into the cell.
- The *output gate* controls whether the updated cell value contributes to the hidden state.

The designation as gates might suggest a 0-or-1 decision on the flow of information. However, these decisions are usually not binary, and rather the extent is controlled to which a value is processed. Therefore layers that perform gate operations use an activation function σ which returns a value between 0 and 1. In the following, we can think of σ as a sigmoid function, although this is not strictly necessary.

Intermediate vectors: The direct update of the hidden vector $\vec{h}_t$ according to (9.3) is replaced by a more complex interaction of four p-dimensional intermediate vectors, which are only needed during the update process. For simplicity, we restrict ourselves in the following to one-dimensional input and drop the vector notation for $d = p = 1$. The intermediate variables f, i, o and $\tilde{C}$ are set up for each time step t using specific weights W, U and biases b:

forget layer :	$f_t = \sigma\left(W_f\, x_t + U_f\, h_{t-1} + b_f\right)$	(9.8)
input layer :	$i_t = \sigma\left(W_i\, x_t + U_i\, h_{t-1} + b_i\right)$	(9.9)
output layer :	$o_t = \sigma\left(W_o\, x_t + U_o\, h_{t-1} + b_o\right)$	(9.10)
cell input layer :	$\tilde{C}_t = \tanh\left(W_c\, x_t + U_c\, h_{t-1} + b_c\right)$	(9.11)

It should be emphasized that the variable f_t, which refers to the forget layer, must be distinguished from the general function f in (9.1). Moreover, the cell state C_t should not be confused with the intermediate variable $\tilde{C}_t$, which has the function of suggesting a suitable input value for the update of the cell state.

Cell state update: The new cell state C_t is then calculated as the weighted average of the previous cell state C_{t-1} and the input value provided by the intermediate variable $\tilde{C}_t$. The coefficients f_t and i_t are between 0 and 1 but otherwise unrelated, yielding

$$C_t = f_t \cdot C_{t-1} + i_t \cdot \tilde{C}_t. \tag{9.12}$$

In other words, the first term of (9.12) selects how much of the previous cell state should be forgotten (complete reset for $f_t = 0$), whereas the second term allows to vary the extent to which the new input is added to the cell state (fully for $i_t = 1$). To circumvent a permanent reset at the beginning of training, the forget biases are usually initialized to ones [56].

Hidden state update: Finally, the hidden state is updated by accessing the refreshed cell state C_t via

$$h_t = o_t \cdot \tanh{(C_t)}. \tag{9.13}$$

The weight factor o_t is again between 0 and 1, indicating the extent of leakage from the cell state into the hidden state.

From (9.12) we see that the partial derivative of C_t with respect to C_{t-1} is f_t, hence the backward gradients for C_t are simply multiplied by the value of the forget layer. Similarly, the partial derivatives of h_t with respect to the weights require only the derivative of the tanh function, since o_t depends only on W_o and U_o. For the latter weights, the additional derivatives of o_t can be expressed by the sigmoid function itself in case a sigmoid is used as activation function.

The rather elaborate LSTM update procedure is represented graphically on the right side of Fig. 9.4. It is instructive to compare these operations to those of a standard RNN, which is depicted on the left side. In both cases, the green block represents the unit labeled 'A' from Fig. 9.2 that we are already familiar with. For the standard RNN cell, the update of the hidden state is calculated according to (9.3), yielding for $d = 1$

$$h_t = \tanh{(W_f\, x_t + U_f\, h_{t-1} + b_f)}.$$

Standard RNN cell   LSTM cell

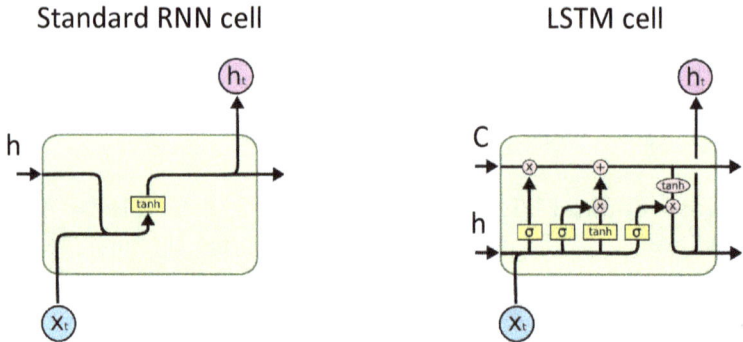

Fig. 9.4. Comparison between a standard RNN cell (left) and a long short-term memory cell (right) [53].

This activation layer is represented in Fig. 9.4 by a yellow block marked 'tanh'. The update of the hidden state is realized along a horizontal line labeled 'h'. It is sometimes compared to a conveyor belt or highway since it transports the hidden states from the left to the right.

By comparison, the LSTM cell contains an additional horizontal line for the cell state labeled 'C' and four activation layers which process h_{t-1} and x_t. It can be interpreted as a shortcut (see Sec. 7.5) through time. Again, these layers are indicated as yellow blocks and marked according to their activation functions. They generate the intermediate variables as specified in (9.8)–(9.11).

Figure 9.5 provides a step-by-step walk through of LSTM operations replicated from the excellent blog [53]. The first step is to use the forget layer to decide to what extent f_t information should be retained by the cell state (upper left in Fig. 9.5). In the second step a candidate value $\tilde{C}_t$ is prepared for the update of the cell state, together with some weight i_t for its subsequent admixing (lower left). The actual update of C_t is executed in the third step, with the upper right of Fig. 9.5 visualizing Eq. (9.12). The hidden state is updated in the last step according to (9.13), depicted on the lower right. This part of the figure clearly shows that the update of the hidden state is based on a constrained, filtered version of the refreshed cell state, as it is passed through an additional activation layer and weighted by a leakage factor o_t.

LSTM architectures have been implemented in many variations in terms of the number and functionality of gates. For example, the initial version of the LSTM did not include a forget layer, introduced in [57]. In other

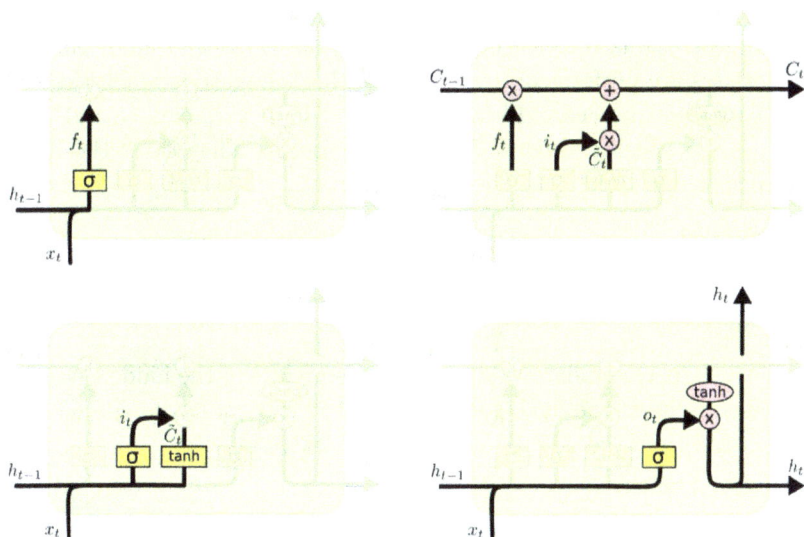

(a) Setup of intermediate variables.　(b) Cell and hidden state updates.

Fig. 9.5. Update process of an LSTM cell explained step by step [53].

cases, the output activation function was omitted [58], or the forget and input gates were coupled to force older data to be forgotten only when new data are learned. Another important variant includes so-called 'peephole connections', which allow the gate layers to take into account the cell state for their decisions [59].

The parameters of a recurrent network with LSTM units are adjusted by training, using a set of training sequences together with an optimization algorithm, e.g. gradient descent. Especially the weights of the gate connections need to be learned during training, since they control how effectively the gates work. The BPTT algorithm, addressed in Sec. 9.3, is employed for the optimization process.

9.5　Gated recurrent unit

Introduced in 2014, the gated recurrent unit (GRU) is a much younger architecture than LSTM [60]. It aims at simplifying the LSTM-update process and is widely used nowadays. GRUs have no cell state and only two gates to control the update of the hidden state at each time step. While the LSTM tightly controls the information passed to the hidden state via separate forget, input, and output gates, the GRU uses only a reset and an update gate, the latter merging the roles of the forget and input gates.

Updates are performed directly on the hidden state for each time step t in a two-step process. First, the intermediate variables r_t and z_t are set up using specific weights W, U and biases b. They enter the further intermediate variable $\tilde{h}_t$ in the following way:

$$\text{reset layer}: \qquad r_t = \sigma\left(W_r\, x_t + U_r\, h_{t-1} + b_r\right) \qquad (9.14)$$

$$\text{update layer}: \qquad z_t = \sigma\left(W_z\, x_t + U_z\, h_{t-1} + b_z\right) \qquad (9.15)$$

$$\text{candidate layer}: \qquad \tilde{h}_t = \tanh\left(W_h\, x_t + U_h\left(r_t \cdot h_{t-1}\right) + b_h\right) \qquad (9.16)$$

Here, $\tilde{h}_t$ is a candidate value used in the update of the hidden state and can be considered the analogue of the LSTM cell input variable $\tilde{C}_t$. This can be seen from the update of the hidden state

$$h_t = (1 - z_t) \cdot h_{t-1} + z_t \cdot \tilde{h}_t, \qquad (9.17)$$

which is analogous to (9.12), considering that for a GRU the roles of the cell state and hidden state are merged. We note that in contrast to (9.12), the coefficients of h_{t-1} and $\tilde{h}_t$ add up to unity. Therefore, new information is stored in the hidden state only at the expense of previous information. The GRU operations defined above are illustrated graphically in Fig. 9.6, similar to Fig. 9.4 for the standard RNN and LSTM cell.

For simplicity, we have again restricted ourselves to a one-dimensional input sequence as in Sec. 9.4. Nevertheless, for higher dimensions the above equations can easily be generalized by replacing the dot by the Hadamard (element-wise) product of vectors.

Both the GRU and LSTM cells described here share the ability to partially reset the hidden state and control the influx of new information. However, the entirely different gating mechanisms indicate that the GRU

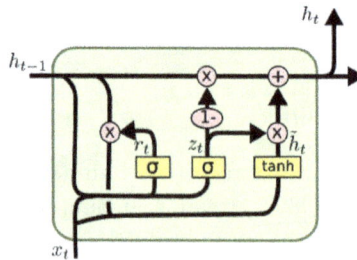

Fig. 9.6. A gated recurrent unit [53].

is *not* a special case of the LSTM. For a detailed comparison, the reader is referred to Ref. [61]. In general, both cell types perform about the same, although comparisons appear to depend strongly on the specific task used for benchmarking. Clearly, GRUs are simpler to implement and have the advantage of requiring fewer parameters, which contributes to their growing popularity. LSTM networks, on the other hand, have been in use for much longer. They are often considered the safer option, especially when long sequences and extensive data sets have to be processed.

Finally, we note that GRUs have also evolved into a variety of types that differ in the way their gates interact with the hidden state or other layers. In particular, the minimal gated unit deserves mentioning, which is an even further reduced unit with only a single forget gate [62]. Not surprisingly, all of these types have their own set of update equations.

9.6 Application areas

Recurrent networks have applications in a broad range of fields. Many of these involve sequential data other than time series, most notably natural language processing. The latter is closely related to semantic parsing, which is the conversion of natural phrases into a machine-interpretable form. This greatly simplifies human-device interaction, with voice assistants being a prime example. Widely used today, they convert speech into text that is then used to retrieve information or perform a predefined task.

However, recurrent networks can be applied to *any* problem where the order of elements generates a context which must be included to generate meaningful output. A good example is the processing of protein data in bioinformatics. Since protein sequences differ in length, RNNs provide the flexibility necessary to deal with this type of variable input [63]. Moreover, while in speech or handwriting recognition the input elements are words, the network output is not limited to producing a typed variant of the input sequence. Therefore, recurrent networks are able to provide abstract forms of output, as exemplified by language translation.

The above examples indicate that the output of RNNs can be quite complex. Since essentially any type of output can be assigned to any type of input, a more systematic approach is required. We therefore go back to (9.1) and recall that the vector $\vec{x}_t$ stands for any type of ordered input, including simple numbers (e.g. temperature readings) as well as high-dimensional arrays (e.g. images with color-coded pixels). It should be emphasized that the output vector $\vec{y}_t$ does not have to be of the same type as the input vector. For example, if we use an RNN to predict the readings of a single temperature sensor, then the input and output sequences both consist of simple numbers. By contrast, if we wish to label a video automatically frame by frame, then the input sequence consists of images (video frames), whereas the output elements are text.

Both cases have in common that each new element $\vec{x}_t$ of the input sequence produces directly one new output element $\vec{y}_t$, which results in synchronized sequences. This is the standard mapping of a RNN, also called (synchronized) 'many-to-many' configuration [see Fig. 9.7(a)]. However, depending on the particular problem, parts of the input or output sequence may not be needed. Table 9.1 provides an overview of various constellations, together with an application example for each case. Consequently, Fig. 9.1 must be expanded to Fig. 9.7 by omitting elements of either the input or the output sequence, or both. For instance, the output sequence may be delayed with respect to the input, as depicted in part (b) of Fig. 9.7. This is typically the case with automatic translations, when at least a few words are needed to create the context on which further translation is based. In extreme cases, a sequence can even consist of only one element. Figure 9.7(c) shows an input sequence reduced to a single element, followed by an output sequence of arbitrary length. An example is the automatic captioning of images, where the input is limited to a single image,

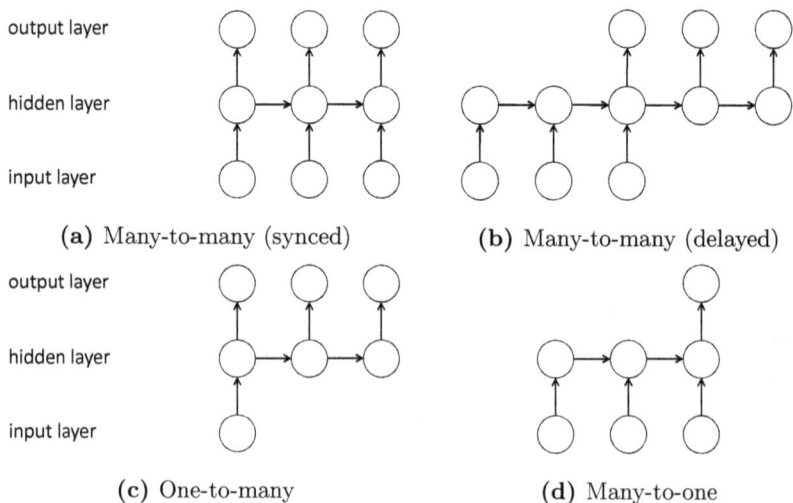

(a) Many-to-many (synced)   (b) Many-to-many (delayed)

(c) One-to-many   (d) Many-to-one

Fig. 9.7. Variants of sequence processing.

Table 9.1. Variants of RNN sequence processing.

Input	Output	Application example	Fig. 9.7
sequence	sequence (synchronized)	temperature forecast	(a)
sequence	sequence (delayed)	automatic translation	(b)
single event	sequence	image captioning	(c)
sequence	single event	text classification	(d)

followed by a text sequence. The reverse case is shown in Fig. 9.7(d), when a single output element is only produced at the end of the input sequence. An application example here would be text classification, e.g. for automatic forwarding of incoming e-mails, where the input is a text sequence that is to be captured in its entirety before any output action is taken. Usually, such *many-to-one* models are constructed by only passing the RNN cell output of the last received time step to the subsequent layer.

Finally, we would like to note that there are many cases of ordered data for which the direction in which the data are processed is not important. Therefore, sequences with dependencies in both directions may also be evaluated using two RNNs tied together as shown in Fig. 9.8. Each RNN operates in one direction, and their output is combined at each step. This so-called *bidirectional* architecture can be particularly helpful when processing ambiguous input data.

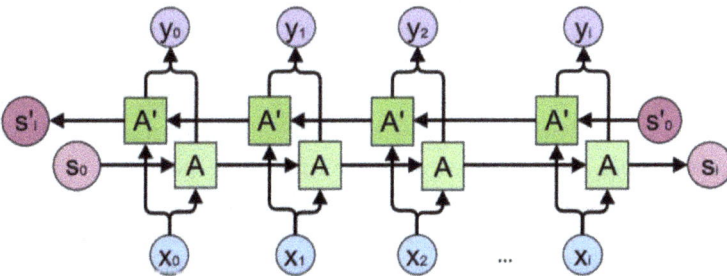

Fig. 9.8. A bidirectional recurrent network [53].

Example 9.1. LHC monitoring: The Large Hadron Collider (LHC) was already mentioned in Example 8.9. Since its superconducting magnets keep the circulating protons in the vacuum tube on the proper trajectory, the stability of these magnets is crucial for successful operation and therefore closely monitored. The suitability of RNNs for fault prediction was investigated in an interesting study [64].

The occasional unintentional loss of superconductivity in a magnet, also called quench, sets free a large amount of energy that can be sufficient to destroy the magnet unless the energy is dissipated in a controlled manner. Quenches appear to be triggered frequently by the release of local mechanical stress, which can result from the device assembly, thermal contraction during cooling, or electromagnetic forces acting on the coils during regular operation. Therefore, the magnets are equipped with an electronic monitoring system that records voltage-time series. When the superconducting state is lost, the resistance becomes finite, and a voltage develops that can be measured and further processed. These data are recorded every 400 ms to monitor performance. Moreover, they are kept for years in a logging system, providing a natural opportunity to train an RNN.

The detailed causes of most quenches are unknown, however minor quench precursors that precede the onset of instability are regularly observed. It should be emphasized that such gradual unfolding in time is the prerequisite for modeling the voltage-times series with a recurrent network. RNNs with 128 LSTM cells were set up and tested in several configurations. The networks were trained with real voltage-time data for a commonly used type of magnet. The data were extracted from a 24-hour time window prior to each of the 425 quench events registered between 2008 and 2016. Despite the low time resolution of the training data, the RNNs proved to be able to predict incipient magnet failure.

The authors also discussed response times for a potential application in real-time control. From a literature review, they concluded that it is possible to lower the time scale for a single iteration to the microsecond range, provided customized hardware and only the internal memory of the processing unit are used.

9.7 Physics applications

Since recurrent networks are especially adapted to nonlinear time series, they can be used for outlier detection and nonlinear dynamics. Hence, our first Example 9.1 will be from particle physics and discuss a monitoring system for the superconducting magnets of the LHC. The second Example 9.2 deals with the important problem of crack growth and propagation in materials physics. In both cases, time series were modeled with RNNs using LSTM cells.

Example 9.2. Crack propagation: Mechanical failure of brittle materials usually occurs catastrophically, with the growth of small cracks that rapidly propagate and coalesce, leading to macroscopic fracture. Accurate prediction of crack growth is a challenging problem in materials science due to the nonlinear coupling of the damage and stress fields involved, as stress accumulation at crack tips causes existing cracks to propagate further and new cracks to form. Moreover, the evolution of the damage field dynamically changes the boundary conditions of the stress field.

Although the equations describing the physics at the microscale are known in principle, it is computationally extremely expensive to simulate crack growth over longer time scales, making it problematic to obtain meaningful predictions at the macroscale.

In Ref. [65], an efficient crack propagation emulator based on recurrent neural networks was introduced. The authors used RNNs with LSTM cells to track the time evolution of stress and damage in a simulated material. However, they also included spatial information, extending the LSTM neural networks to a spatio-temporal model. In addition, they took advantage of the differential equations governing this problem by including partial derivatives from the stress field in the input to improve predictions. They also modified the loss function to take into account certain quantities of interest, such as the number of cracks and their length distribution.

It was found that this so-called *physics-informed spatio-temporal* approach faithfully reproduced both stress and damage field propagation, allowing the prediction of summary statistics for quantities of interest from initial conditions to the failure point.

9.8   Exercises

Exercise 9.1. Sinus-wave (forecasting)

(1) Run the provided script (Exercise 9.1 from the exercise page) and comment on the output.

(2) Vary the number and size of the LSTM layers and compare training time and stability of the performance.

Summary 9.1.

(1) Recurrent neural networks (RNNs) are a special type of artificial neural network suitable for sequential data such as time series, text, and other ordered phenomena. They are distinguished by an additional internal memory that allows to process input sequences of variable length.

(2) RNN building blocks are called cells. They can be symbolized by a self-loop, which represents the update of the hidden state after each input. For a series of finite length, the loop can be unrolled into a layered network.

(3) RNNs are trained by a backpropagation algorithm adapted to sequential data, called backpropagation through time.

(4) The classical realization of an RNN cell is the long short-term memory (LSTM). LSTM cells have a further internal memory called cell state, which is actively controlled by gates and interacts with the hidden state.

(5) Another important realization of an RNN cell is the gated recurrent unit (GRU). GRUs have no cell state and only two gates to control the update of the hidden state.

(6) The output sequence of an RNN does not have to be of the same type as the input sequence.

(7) The output sequence can be synchronized or delayed with respect to the input sequence. Moreover, elements of either sequence may be omitted. In particular, an output may be generated only after the entire input sequence has been processed.

Chapter 10

Graph networks and convolutions beyond Euclidean domains

Scope 10.1. This chapter presents the concept of graph networks, which allow us to analyze data distributed on non-Euclidean domains. In this chapter, we first review the basics of graph theory. We then introduce convolutions on graphs using a realization of filters in the spatial and spectral domains. We further discuss several approaches and network architectures suited for various data structures and applicable to diverse physics challenges.

10.1 Beyond Cartesian data structures

While many detectors feature Cartesian sensor placements, often non-Cartesian or even non-regular arrangements of sensors are used. Additionally, some experiments provide data that lie on a non-Euclidean manifold, like a sphere (compare Fig. 10.1).

Extending the convolutional operation, as introduced in Sec. 8.1, to irregular, i.e. non-regular or even non-Euclidean grids, is not straightforward. The properties and strengths of a convolutional operation are:

- translation invariance,
- scale separation (learning features of different hierarchies),
- number of parameters is independent of the input,
- and deformation stability (the filters are localized in space).

Mainly, the last two points cause problems. It is rather challenging to define a filter with a number of adaptive parameters independent from the input size in a space that is neither regular nor Euclidean. To define a convolution

(a)

(b)

30 GHz

(c)

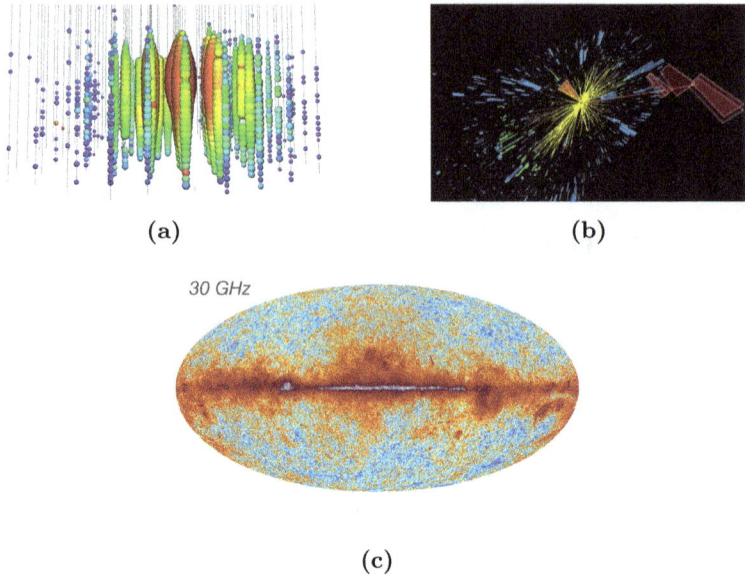

Fig. 10.1. Data structures encountered in physics experiments. (a) Non-Cartesian: data measured by the IceCube experiment. The experiment features strings of 60 optical modules each, arranged in a hexagonal grid [66] (*©reprinted with permission of the IceCube Collaboration*). (b) Non-regular: measurement data of the CMS experiment. The experiment highlights various detector modules with different sensor arrangements [67] (*©CERN, permission for educational use*). (c) Non-Euclidean: spherical radiation data measured by the Planck spacecraft. (*Credit: Planck Collaboration [68], reproduced with permission ©ESO.*)

on such data, it is helpful to understand the underlying structure as a graph. Using graph convolutional neural networks, based on basic graph theory, it is possible to exploit the data on irregular domains.

First graph neural networks were introduced several years ago [69, 70]. Inspired by these, recent works focus on extending graph networks by utilizing convolutions beyond Cartesian domains. To implement these, several extensions to well-known deep learning frameworks exist. Using the Spektral library [71] graph networks can be implemented in Keras and Tensor-Flow. PyTorch users can make use of the extensive PyTorch Geometric [72] extension. An alternative is the Deep Graph Library [73] which can be used on top of TensorFlow, PyTorch, and MXNet.

Approaches allowing graph networks to analyze data lying on complex manifolds using convolutions can be divided into two broad categories. On the one hand, networks utilize convolutions in the spatial domain by generalizing the convolution operation used in CNNs. On the other hand, spectral methods perform convolutions in the Fourier (spectral) domain. In the following, we give a short introduction to graph theory and subsequently discuss spatial followed by spectral methods.

10.2 Graphs

A graph is a pair of nodes $\mathcal{V}$ and edges $\mathcal{E}$, for example representing the neighborhood relation of a cloud of points. One particular property of such a basic graph is that the nodes have no internal order and are therefore invariant under node permutation. Consequently, the representation of a graph is not unambiguous, as shown in Fig. 10.2(a). Several algorithms exist to find meaningful representations of graphs [74]. For plotting graphs, for example, the python package networkx [75] can be used. It provides various representations, and supports a variety of formats.

In the scope of this chapter, we distinguish between three different types of connected graphs:

- undirected graphs,
- directed graphs,
- and embedded (weighted) graphs,

where the latter can be undirected or directed. The different types of graphs are shown in Fig. 10.2. Undirected and unweighted graphs are the most simple graph types and are only defined via an existing or non-existing neighborhood relation. Examples of such graphs are simple social networks. Directed graphs consider further a specific direction of the edges. Additionally, each edge can hold a particular weight. Such graphs are well known as representations of stochastic matrices.

We further consider graphs embedded in a specific manifold, as such data structures are common for physics. This means that the nodes lie in a 'fixed' neighborhood described by at least one property — for example, a non-regular arrangement of sensors that together form a spherical detector. The edges then could describe the geometrical distance between the sensors. In general, their relative positions encode essential information of the underlying structure, which should be taken into account by the

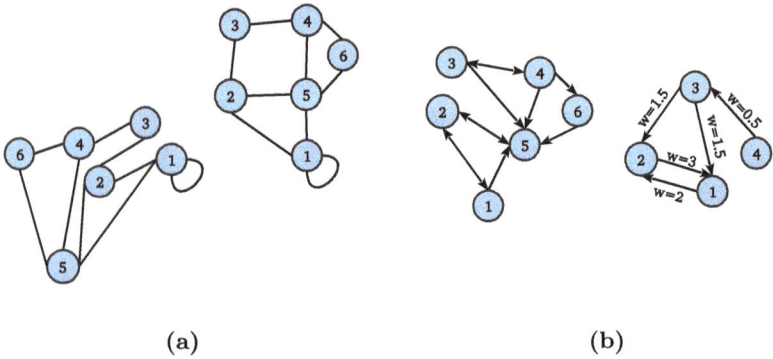

(a)   (b)

Fig. 10.2. Different types of graph. The numbers represent the assigned index for a given node. (a) Two representations of the same undirected graph. (b, left) Directed and (b, right) weighted directed graph.

architectures of graph networks. Other examples are Feynman diagrams. Here, the edges indicate the type of particle (e.g. electron, positron, or photon), and the nodes hold the feature of vertex factors, which describe the coupling strength of the interaction.

10.2.1 Adjacency matrix and graph basics

The relationship of n nodes in a graph is described by the *adjacency matrix* $\mathbf{A}$ of shape $n \times n$. This is similar to an un-normalized version of a stochastic matrix, indicating the transition probability from one state to another. Here, the weight of the edge going from node i to node j is given in row i and column j. See Example 10.1 for the adjacency matrix of the unweighted and undirected graph depicted in Fig. 10.2(a). Since the adjacency matrix covers the graph's structure, it forms the core of defining operations on graphs. Note that undirected graphs feature a symmetric adjacency matrix, as each edge connecting node i and j can be interpreted as an edge $i \to j$ and $j \to i$ with the same weight.

Another interesting property is the kth power of $\mathbf{A}$. Here each element i, j of $\mathbf{A}^k$ corresponds to the number of walks of length k from node $i \to j$. Thus we can express with $\mathbf{A}^k$ the so-called k-hop neighborhood.

A further important matrix in graph theory is the *degree matrix* $\mathbf{D}$, which is used to describe the connectivity of the nodes. For each particular node using $\mathbf{D}$, the number of edges (degree) of a specific node is stated.

Thus, **D** is a diagonal matrix. An example degree matrix is also discussed in Example 10.1.

Example 10.1. Adjacency, degree matrix: Adjacency and degree matrix for the example of an undirected and unweighted graph is shown in Fig. 10.2(a). In the adjacency matrix **A**, each column and row describes the neighborhood relations of a specific node.

$$\mathbf{A} = \begin{pmatrix} 2 & 1 & 0 & 0 & 1 & 0 \\ 1 & 0 & 1 & 0 & 1 & 0 \\ 0 & 1 & 0 & 1 & 0 & 0 \\ 0 & 0 & 1 & 0 & 1 & 1 \\ 1 & 1 & 0 & 1 & 0 & 1 \\ 0 & 0 & 0 & 1 & 1 & 0 \end{pmatrix}, \qquad \mathbf{D} = \begin{pmatrix} 4 & 0 & 0 & 0 & 0 & 0 \\ 0 & 3 & 0 & 0 & 0 & 0 \\ 0 & 0 & 2 & 0 & 0 & 0 \\ 0 & 0 & 0 & 3 & 0 & 0 \\ 0 & 0 & 0 & 0 & 4 & 0 \\ 0 & 0 & 0 & 0 & 0 & 2 \end{pmatrix}$$

For example, node '1' — look in the first row and check the column for outgoing edges of node '1' — is connected to node '2', node '5' and itself via self-coupling (here counted twice). By choosing a respective column and checking the rows, the edges that end at a given node can be identified. Note that in principle, the nodes are invariant under permutation; thus, an arbitrary numbering can be used, and the rows and columns would change. As this example is for an undirected graph, the matrix is symmetric. It features discrete values as the example is an unweighted graph. The degree matrix **D** indicates the number of edges at the particular node and always has a diagonal shape.

Graph Laplacian: Using the degree matrix and the adjacency matrix, the graph Laplacian $\mathbf{L}' = \mathbf{D} - \mathbf{A}$ can be constructed. The Laplacian corresponds to the discrete version of the Laplace operator Δ we know, for example, from Poisson's equation and electrodynamics. To imagine the transition of $\mathbf{L}' \to \Delta$, think of a graph as a mesh (of small triangles [76]) which becomes increasingly dense and converges towards a manifold where the Laplace operator is defined. Using the Dirichlet energy, the Laplace operator can be understood as a measure of smoothness of a scalar function on this manifold [76]. Hence, in a figurative sense, the Laplacian describes how smoothly a signal can propagate across the graph.

In the following we will use the *symmetric normalized Laplacian* defined as $\mathbf{L} = \mathbf{D}^{-\frac{1}{2}}\mathbf{L}'\mathbf{D}^{-\frac{1}{2}} = \mathbb{1} - \mathbf{D}^{-\frac{1}{2}}\mathbf{A}\mathbf{D}^{-\frac{1}{2}}$. This translates to the Laplacian:

$$\mathbf{L}_{ij} = \begin{cases} 1 & \text{if } i = j \text{ and } \deg(v_i) \neq 0 \\ -\dfrac{1}{\sqrt{\deg(v_i)\deg(v_j)}} & \text{if } i \neq j \text{ and node } v_i \text{ is adjacent to node } v_j \\ 0 & \text{otherwise} \end{cases}$$

(10.1)

which further normalizes the adjacency matrix using the number of outgoing and incoming edges v_i, v_j of connecting nodes. Here, $\deg(v_i)$ denotes the degree of node i.

An interesting property of the Laplacian is that its eigenfunctions are also a Fourier basis of the respective graph or manifold. This results from the fact that the Fourier transformation diagonalizes the Laplacian $\mathcal{F}(\Delta f) = -\omega^2 \mathcal{F}(f)$. Remembering the Laplacian eigenfunctions in a 1D Euclidean space $\Delta e^{i\omega x} = -\omega^2 e^{i\omega x}$ illustrates the connection to the Fourier transformation even further. Hence, the eigenvalues of the graph Laplacian can be understood as frequencies and its eigenfunctions as Fourier modes of the graph.

Before using this link to extend the convolutional operation to the spectral domain in Sec. 10.5.1, we first discuss approaches utilizing the spatial domain.

Node features: In addition to the edge features described by the adjacency, graphs can further feature node features $\vec{v}$. These features can describe properties at the given node. For a social network, the features could correspond, for example, to the gender, job, and age of a person. In graphs constructed for physics experiments, the node features usually indicate measurements performed at the respective location or properties of a particular particle or event in a point cloud.

10.3 Construction of graphs

In computer science, the analysis of data that naturally lie on graph structures is often of interest. These data sets, e.g. social networks or citation networks, already have a graph-like structure that can be directly used to determine an adjacency matrix and a Laplacian. In contrast, most physics data sets do not feature such structures but lie on complex[1] manifolds.

---

[1] Here, complex means that the manifolds have a non-trivial structure, not necessarily that imaginary numbers are involved.

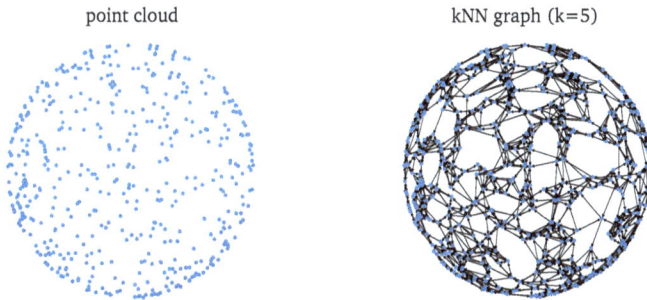

point cloud   kNN graph (k=5)

Fig. 10.3. Construction of a graph on a point cloud. Using k-nearest neighbor clustering with $k = 5$, the neighborhood relations are defined for each point.

These data can often be interpreted as a point cloud, i.e. points embedded in a multi-dimensional space (compare with Fig. 10.1).

To obtain a graph-like structure, clustering operations can be used. Assuming a spherical detector with an irregular placement of sensors with given coordinates, we can use a clustering operation to define the neighborhood relations for every single sensor. For example, we can use for each sensor a kNN (k-nearest neighbors) classification by making use of the sensor coordinates. Choosing $k = 5$, each sensor would have five neighbors, corresponding to a node with 5 edges. An illustration of a formed graph from a spherical point cloud using kNN classification is shown in Fig. 10.3. Note that due to the kNN operation, the obtained graph is directed.

10.4 Convolutions in the spatial domain

Below, we introduce methods that perform the convolution in the spatial domain, i.e. by performing filtering operations with filters localized in space.

10.4.1 *Graph convolutional networks*

The most intuitive way to design a graph convolutional network would be a direct transfer of the convolutional operation from Sec. 8.1 to graphs. Representing image-like data as a graph is straightforward. In Fig. 10.4(a), an example graph is shown, where each node is connected to its eight closest neighbors (neglecting the edges here). In a standard CNN we now would have for a 3×3 kernel applied to node (pixel) i, nine independent weights

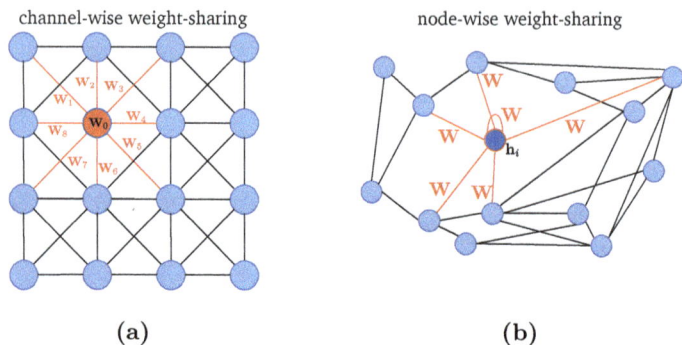

Fig. 10.4. Comparison between the convolutional operation on (a) image-like data (regular grid of pixels) and (b) connected undirected graphs as utilized by graph convolutional networks (GCNs).

$W_0, ..., W_8$ per dimension of the input $h_i^{(0)}$. For a color image, the input dimension would be three because of the three color channels.

Since for image-like data, the underlying pixel grid is regular, the weights can be shared over the image. This is not feasible for a non-regular grid, shown in Fig. 10.4 as the number of neighbors and the distance between the nodes can change from place to place. Therefore, transferring the convolutional operation to graphs requires some modifications.

The simplest approach is a node-wise weight-sharing illustrated in Fig. 10.4(b). Here, each node holds a feature vector $\mathbf{h}_i^{(0)}$, and the same weight matrix $\mathbf{W}$ is applied to each node. When applied to a central node, only the information of the neighboring nodes $j \in \mathcal{N}_i$ and the central node itself (self-coupling) are considered in the convolutional operation.

When aggregating the contributions of the neighboring nodes using addition and applying a nonlinearity σ, the operation at node i reads:

$$\mathbf{h}_i^{\prime\top} = \sigma\left(\mathbf{h}_i^\top \mathbf{W} + \sum_{j \in \mathcal{N}_i} \frac{1}{c_{ij}} \mathbf{h}_j^\top \mathbf{W}\right) \qquad (10.2)$$

where $\mathbf{h}_i^{\prime\top}$ denotes the row vector of the node features and c_{ij} are constants for normalization purposes. Otherwise, the produced activation at a node would depend on the number of its edges.

The simple model of (10.2) can be formulated as a general propagation rule for graph convolutional networks (GCNs) [77]. Let us assume a data structure that can be modeled using a graph with n nodes and a known adjacency matrix $\mathbf{A}$ (dimension $n \times n$) that describes the graph. For an input

X with dimensionality $n \times d$ — each node holds d features — the weight matrix **W** would consist of $d \times f$ weights. This defines the convolution to be independent of the number of nodes, as for each node d inputs are transformed into f features. Thus, the propagation would read $\mathbf{H} = \mathbf{XW}$.

To consider the structure of the graph in the operation, the adjacency matrix of the graph needs to be included since **A** models the relationships between the nodes. Here, **A** corresponds to $j \in \mathcal{N}_i$ used in (10.2) as it gives for each node i all neighboring nodes. Note that in the GCN model self-coupling is used, so that the adjacency for the operation is:

$$\hat{\mathbf{A}} = \mathbb{1} + \mathbf{A}$$

As the adjacency matrix needs to be normalized to preserve the feature scale, the degree matrix $\hat{\mathbf{D}}$ of $\hat{\mathbf{A}}$ is used. By using a symmetric normalization

$$\hat{\mathbf{A}} \to \hat{\mathbf{D}}^{-\frac{1}{2}} \hat{\mathbf{A}} \hat{\mathbf{D}}^{-\frac{1}{2}},$$

the adjacency is inversely scaled by the number of edges at each node (replacing the normalization constants c_{ij} in (10.2), which is similar to the normalization in (10.1)). Finally, the propagation rule for a GCN layer reads:

$$f(\mathbf{H}, \mathbf{A}) = \sigma\left(\hat{\mathbf{D}}^{-\frac{1}{2}} \hat{\mathbf{A}} \hat{\mathbf{D}}^{-\frac{1}{2}} \mathbf{HW}\right) \tag{10.3}$$

Stacking several of these GCN layers, which all share the same graph structure (the adjacency matrix remains the same), the structured data can be exploited.

To illustrate the GCN operation and clarify the similarities and differences to CNNs, a 1D application is shown in Fig. 10.5. Here, the shapes of the matrices are depicted explicitly. This comparison is similar to the example discussed in Sec. 8.2. Here, as an example, a time series of five numbers is assumed. It is suitable for 1D convolutions and can be interpreted as an undirected and unweighted graph. Each node holds a feature vector $\vec{f}$ of the same dimension (three 'channels' in this case). It can be easily identified that the adjacency matrix has a similar structure as the weight matrix in the standard CNN case (compare Fig. 8.5).

GCNs were initially proposed for semi-supervised classification tasks (see Example 10.2), but together with recent developments [78–80] architectures based on GCNs are applied in physics and other sciences [81, 82]. Although the GCN algorithm has linear complexity and can be well understood as aggregating the neighborhood information, it is limited to undirected graphs and scalar edge features.

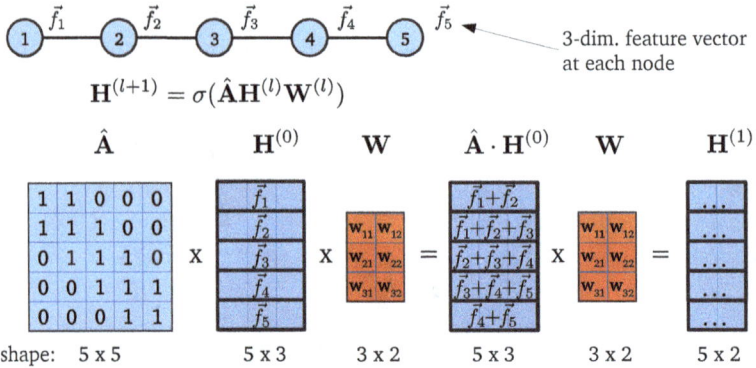

Fig. 10.5. Detailed visualization of the GCN propagation rule. On a graph with five nodes with three features each, a GCN layer is applied with two filters.

Example 10.2. Zachary's karate club: Semi-supervised classification on a social network (Zachary's karate club [83]). After a dispute, a karate club consisting of 34 people splits into four groups. Only one person of each group is known. Hence, the task is related to incomplete supervision (see Sec. 16.1). Besides, a social network is known and indicates close social relations. To assign each karate enthusiast (node) the correct affiliation, a GCN can be used to analyze the social network structure: an undirected, unweighted graph.

As the first step, the normalized adjacency matrix of the graph is to be estimated. In this scenario, the GCN consists of four layers that share the graph structure via the adjacency matrix. The GCN is trained in a semi-supervised manner. During training, only four nodes are labeled (only these colored nodes contribute to the loss, the rest is masked out). An illustration of the example application is shown in Fig. 10.6. After 300 iterations (facing the same data and graph), the predictions for the whole graph can be estimated, which show only a few errors.

Code and additional visualization are available as Exercise 16.1.

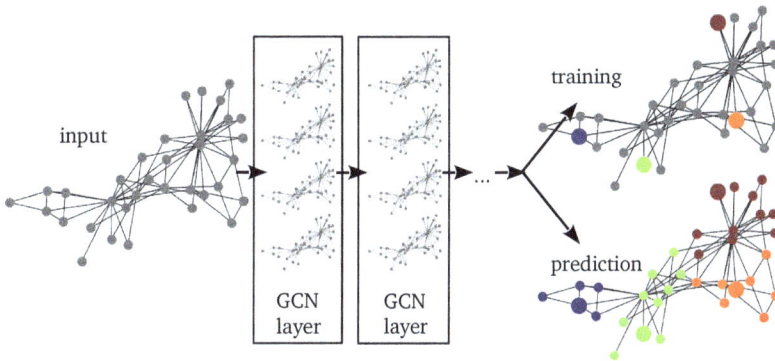

Fig. 10.6. Application of a GCN to a social network: Zachary's karate club [83]. Nodes denote club members and edges social relations. The network is trained in a semi-supervised manner to predict the membership of individual club members.

10.4.2 *Edge convolutions*

A second well-known approach for graph convolutions in the spatial domain is the edge convolution method [84, 85]. One significant advantage of edge convolutions is that they do not require a graph but rather a point cloud, i.e. points embedded in a d-dimensional space. For performing the convolution, a graph is formed defining the neighborhood for each point of the cloud. In principle, in each layer new neighborhood relations can be defined. Architectures featuring such updates are discussed in Sec. 10.4.3.

The basic edge convolution can be decomposed into three operations:

(1) the definition of a graph,
(2) the formation of edge features by convolving with a kernel function,
(3) and the aggregation over the neighborhood.

In the first step, out of the point cloud, a graph is formed using k-nearest neighbor (kNN) clustering. For each point $x_i \in \mathbb{R}^M$ (or now rather node) the closest k neighbors x_{i_j} are searched. Note that kNN clustering results in a *directed* graph (non-symmetric $\mathbf{A}$), as for a point i which has the k-closest neighbor j, it is not guaranteed that i is, in turn, one of the k-closest neighbors of j. In Fig. 10.7, the different steps of performing a single edge convolution to the node x_i are shown. Here, $k = 5$ is used as the number of neighboring nodes.

point cloud

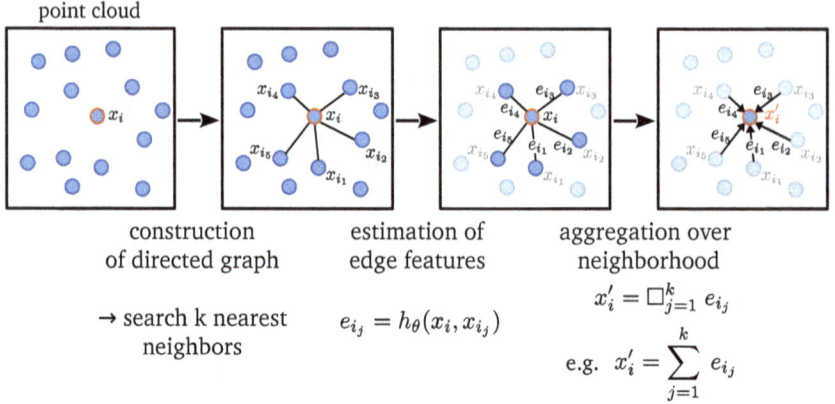

construction of directed graph	estimation of edge features	aggregation over neighborhood

$$x_i' = \Box_{j=1}^{k} \, e_{ij}$$

→ search k nearest neighbors $\quad e_{ij} = h_\theta(x_i, x_{ij})$

e.g. $x_i' = \sum_{j=1}^{k} e_{ij}$

Fig. 10.7. Working principle of the edge convolution on a point cloud, shown for a single point x_i and $k = 5$.

In the second step, edge features are formed by applying the kernel function h_θ to each node x_i. Critically, the kernel function is a neural network holding adaptive parameters θ. Like for CNNs, in the edge convolution, weight-sharing is used as well, i.e. the same kernel function with the same parameters is used for every node. The number of parameters used in the edge convolution is therefore independent of the number of inputs.

The application of the kernel(-function) to the neighborhood is as follows. At node x_i, for all its k neighbors x_{ij}, the so-called edge features e_{ij} are calculated via

$$e_{ij} = h_\theta(x_i, x_{ij}). \tag{10.4}$$

This can be understood as a generalization of a discrete filter used in a CNN, as in general, the kernel function directly depends on the coordinates x_i, x_{ij} of the nodes. The number of new edge features k produced by the continuous kernel is thereby, to some extent, similar to the kernel size used in a CNN. Instead of estimating the filter response by weighting the neighborhood of the central pixel (node x_i) with a discrete filter, a continuous kernel function that depends on the coordinates of the neighborhood nodes is used. This transition from a discrete kernel — as used in a CNN — to a continuous kernel used in the edge convolution is shown in Fig. 10.8 for a 1D example.

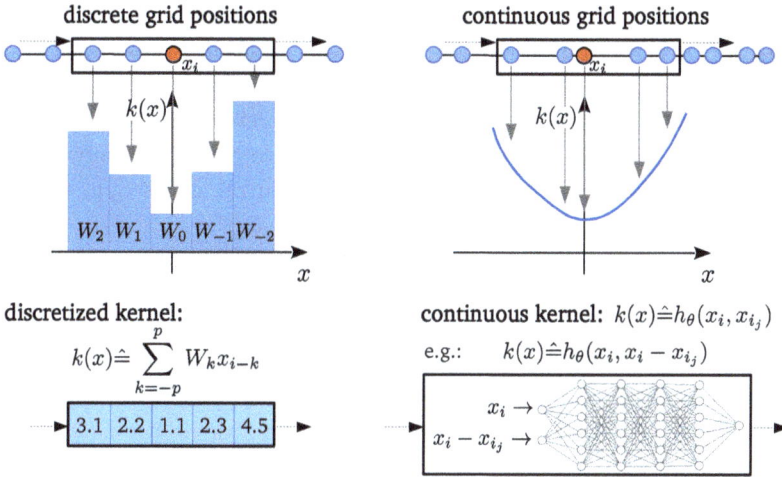

Fig. 10.8. Comparison between a discrete kernel applied to discrete grid positions (left) as utilized in CNNs (here 1D convolution) and a kernel function (right), realized by a neural network h_θ, which can operate on a continuous grid.

The final response of the filter is calculated by performing

$$x'_i = \square_j \, e_{i_j} \tag{10.5}$$
$$= \square_j \, h_\theta(x_i, x_{i_j}), \tag{10.6}$$

an aggregation operation over the neighborhood, making the output size of the edge convolution independent from the number of inputs. Here, $\square_j$ denotes the aggregation operation to be defined by the user, e.g. by summing up $(\sum_j)$ as used in CNNs or using the maximum $\max_j$ as used in pooling operations. The dimension of the resulting feature vector (number of produced features) is controlled by the output dimension of the kernel function (see Example 10.3). Thus, the output dimension could be interpreted as the number of filters in a CNN.

The ability to choose an almost arbitrary aggregation operation and kernel function allows the user to design convolutions with very different properties. This can be particularly helpful to exploit specific symmetries in data. For example, the following design considerations for the edge convolution can be made to ensure specific invariances:

Example 10.3. Kernel network: Assume we want to design a kernel network or kernel function $h_\theta(x_i, x_{i_j})$ given an input of 100 points embedded in a 3D space with coordinates u, v, w. After defining a graph using kNN clustering (e.g. $k = 5$), we want to perform an edge convolution which gives ten new features. The input to the kernel network are six numbers: three for the coordinates of the central node (u_i, v_i, w_i) and three for the respective neighbor node $(u_{i_j}, v_{i_j}, w_{i_j})$.

To obtain ten new features in the end, we can simply design an arbitrary deep neural network — our kernel function h_θ — with six input nodes and ten output nodes. This network is then applied five times to the central node and one neighboring node x_{i_j}, respectively. This results in five ten-dimensional vectors, which are further processed by the aggregation operation, here the average, to obtain a single ten-dimensional vector x_i'. After applying this convolution to each node, the input of 100×3 is changed to 100×10. Note that these convolutions are performed for all 100 points in parallel and do not contain information of the x_i' constructed by neighboring operations.

- if $h_\theta(x_i - x_{i_j})$ — translational invariance,
- if $h_\theta(x_i, x_i - x_{i_j})$ — translational invariance + local information,
- if $h_\theta\left(|x_i - x_{i_j}|\right)$ — rotational invariance,
- if $\square$ is a symmetric operation — permutational invariance.

The last property represents the independence of the result on the order of nodes, which is a distinct feature of graphs.

According to symmetries and additional information in the data, the convolution can be further extended by not only making use of the position but by including additional properties of these points as well. See Example 10.4 for more details.

10.4.3 *Dynamic graph update*

When stacking several layers of edge convolutions on top of each other, the same graph can be used to define the neighborhood. But it can be useful to rebuild the graph in every new layer. Such updates of the underlying graph are utilized in dynamic graph convolutional neural networks (DGCNNs) [85, 87].

Example 10.4. Extension of edge convolutions: In principle, an edge convolution-based graph network is not restricted to using node co-ordinates (position of the points, used to form the graph) only. Assume we measured additionally the properties f_i of the particles at coordinates x_i. Now, we could design an edge convolution which defines the graph using x_i to exploit the spatial data structure. But the kernel function could depend on the features as well $h_\theta = h_\theta(x_i, x_{i_j}, f_i, f_{i_j})$ or e.g. $h_\theta(x_i, x_i - x_{i_j}, f_i, f_{i_j})$. Combining the inputs in that kernel network could then be arbitrarily implemented, e.g. to apply a specific transformation before mixing the features with the coordinates. Knowledge of the data and the underlying symmetries can be decisive here.

Instead of using a single k-nearest neighbor clustering (kNN) to form a graph $\mathcal{G}$ out of the input point cloud in the first layer, in DGCNNs, a new graph is formed in each subsequent layer. Here, in each layer kNN clustering is used to form a new graph $\mathcal{G}'$ using the obtained features x_i'. In other words, this dynamic updating of the graph allows the network to find new neighborhoods. Thus, after the first layer, the convolution is not defined by considering neighbors in the spatial domain but neighbors in the feature space. Here, close distances do not correspond to spatial proximity but similarity. The principle of the dynamic graph update that results in new neighborhoods is depicted in Fig. 10.9(a).

It is important to stress that here the finding of new neighborhoods differs from 'real' learning. kNN clustering features no gradient, so the gradient cannot be backpropagated through the clustering operation. Still, it is to be expected that related points are embedded relatively close in the feature space.

Even though the algorithm can suffer from disregarding local correlations in the spatial domain, it can be powerful for setups with only a few known symmetries. Here, a clustering of neighborhoods beyond the prior embedding input to the network can be exploited, which allows for a convolution that convolves nodes not determined by local proximity but by similarity.

In fundamental physics, several approaches exist which utilize DGCNNs. Examples are networks designed for the application to calorimeters [88], the identification of cosmic-ray sources [89], or jet tagging [86]. The architecture of the so-called ParticleNet, designed for the tagging of particle jets, is illustrated in Fig. 10.9(b). The particular definition of edge convolution

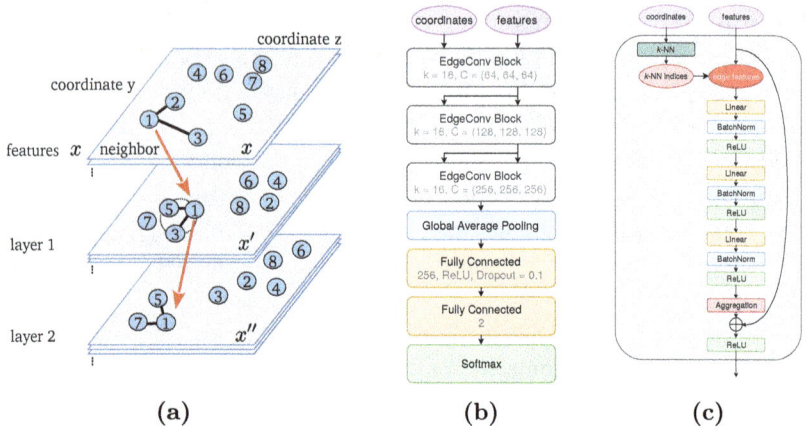

Fig. 10.9. (a) Dynamic graph update as used in DGCNNs. In each layer a new graph is defined based on the neighborhood relation of the coordinates generated by the output of the previous layer. (b) Layout of ParticleNet [86], a DGCNN architecture designed for tagging of particle jets. (c) Design of the edge convolution used in ParticleNet.

and the definition of the kernel function are shown in Fig. 10.9(c). The kernel function features three layers, batch normalization, and a residual shortcut. ParticleNet achieved the highest accuracy in a popular jet tagging benchmark [90].

10.4.4 *Deep Sets*

Finally, let us briefly consider an even more simplified approach: If we imagine all nodes not to be connected to one central node, but not to each other, we have reproduced so-called *Deep Sets* [91,92]. While less expressive than a complete graph network, deep sets still offer a relatively simple to implement and computationally cheap approach to work with unordered inputs.

10.4.5 *Interpretation using the message passing neural network framework*

Using the framework of message passing neural networks (MPNNs), proposed by Ref. [82], a wide class of graph networks can be formulated. This framework divides the network operation into three subsequent phases:

(1) message phase,
(2) aggregation phase,
(3) update phase,

that are simultaneously performed for each node. This is similar to the steps defined in the previously discussed edge convolutions (see Sec. 10.4.2) and can be interpreted as a more general concept.

Message phase: In the *message phase*, each node i collects messages $\vec{m}_{ij}$ from its neighboring nodes $j \in \mathcal{N}_i$. The message function f resulting in messages $\vec{m}_{ij} = f(\vec{x}_i, \vec{x}_j, \vec{e}_{ij})$ can depend on the node features $\vec{x}_i, \vec{x}_j$ and edge features $\vec{e}_{ij}$.

Aggregation phase: In the following *aggregation phase*, using a operation (usually invariant under permutations) for each node i, the received messages are aggregated to a single message $\vec{m}'_i = \Box_{j \in \mathcal{N}_i} \vec{m}_{ij}$.

Update phase: In the final step, for each node a new feature vector is obtained by performing the *update step* using $\vec{x}'_i = g(\vec{x}_i, \vec{m}'_i)$. The function g often depends on the old features, which can be interpreted as a form of self-coupling.

Example 10.5. Message passing network: Using the MPNN framework the GCN operation rule can be interpreted as follows. The message phase is $\vec{m}_{ij} = \frac{1}{\sqrt{\deg(i)\deg(j)}} A_{ij} \vec{x}_j$, as each node receives messages according to the normalized adjacency matrix (which includes self-connections). Then, the received messages are summed up. Thus, the aggregation operation is given by $\vec{m}_i = \sum_{j \in \mathcal{N}_i} \vec{m}_{ij}$. The final update phase is $\vec{x}'_i = \sigma(\mathbf{W}\vec{m}_i)$, where $\mathbf{W}$ is the weight matrix and σ an activation to be chosen by the user.

10.5 Convolutions in the spectral domain

Instead of utilizing the convolutional operation in the spatial domain, i.e. by defining local filters to connect neighboring nodes, convolutions on graphs can be realized in the Fourier domain [78, 93], also called spectral domain. Mathematically, a convolution in the spatial domain with a filter g acting on f is defined as:

$$(f * g)(x) := \int_{\mathcal{R}^n} f(\tau)g(x - \tau)\mathrm{d}\tau. \qquad (10.7)$$

By transforming to the Fourier (spectral) domain via $\hat{f} = \mathcal{F}\{f\}$, due to the *convolution theorem* [94], the convolution in the spatial domain can be rewritten as pointwise operation in the Fourier domain $\mathcal{F}\{f * g\} = \hat{f} \cdot \hat{g} = \hat{g} \cdot \hat{f}$. Therefore, it can be beneficial to rewrite the convolutional operation using the Fourier transform

$$f * g = \mathcal{F}^{-1}\left\{\hat{g}\hat{f}\right\}, \qquad (10.8)$$

with $\hat{f}, \hat{g}$ defined in the spectral domain.

In the (discrete) case of graphs, the Fourier transformation is done with the orthonormal eigenvectors, also called graph Fourier modes, of the graph Laplacian. They form a Fourier basis of the respective graph[2] with

$$\mathbf{L} = \mathbf{\Phi}\mathbf{\Lambda}\mathbf{\Phi}^T.$$

Here, $\mathbf{\Phi}^T$ is the Fourier transformation and $\mathbf{\Phi} = (\phi_0, ..., \phi_n)$ its inverse and contains the Laplacian eigenvectors (ϕ_i). The diagonal matrix $\mathbf{\Lambda} = \mathrm{diag}(\lambda_1, ..., \lambda_n)$ is the matrix of eigenvalues and further illustrates the naming of the Fourier domain as spectral domain.

In Fig. 10.10 an example for the eigendecomposition of a graph is shown. Here for an image-like graph (of dimension 28×28), its 20 first eigenvectors are shown. The associated eigenvalues λ_i of the Laplacian can be interpreted as eigenfrequencies.

This decomposition is similar to a principal component analysis (PCA) obtained using a PCA of the graph. And in fact, the results are inversely proportional[3] to the eigenvalues of the graph [96].

10.5.1   *Spectral convolutional networks*

Spectral convolutional networks use the convolution theorem to realize the convolution in the spectral domain. Assume we want to analyze an incoming signal f using a filter w. In matrix-vector notation this translates to $f * w = \mathbf{W}_\theta f$, where now $\mathbf{W}_\theta$ corresponds to a weight matrix in the spatial domain with adaptive parameters. In contrast to this spatial formulation, we now want to learn filters in the spectral domain. Therefore, we need to use the Fourier transform to transform a signal f to the spectral domain

---

[2]See Sec. 10.2.1.

[3]Due to the fact that the Laplacian features a minus sign. For more details, refer to [95].

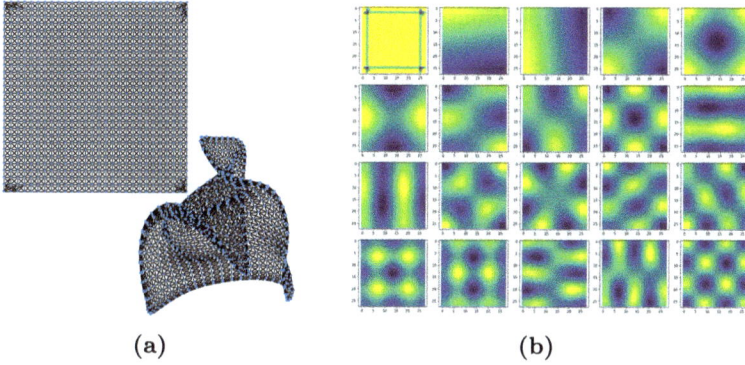

(a)   **(b)**

Fig. 10.10. Eigendecomposition of an image-like graph $\mathcal{G}$. (a) Two representations of $\mathcal{G}$ and (b) the first 20 of ($28 \times 28 =$) 784 eigenvectors of its graph Laplacian.

($\hat{f} = \mathbf{\Phi}^\mathbf{T} f$). That will simplify the convolution with the filter, as filters act pointwise in the spectral domain. The convolutional operation according to (10.5) reads:

$$f' = f * w = \mathbf{W}_\theta(\mathbf{L})f = \mathbf{W}_\theta(\mathbf{\Phi}\mathbf{\Lambda}\mathbf{\Phi}^T)f = \mathbf{\Phi}\hat{\mathbf{W}}_\theta(\mathbf{\Lambda})\mathbf{\Phi}^T f = \mathbf{\Phi}\hat{\mathbf{W}}_\theta\mathbf{\Phi}^T f,$$
$$(10.9)$$

where $\hat{\mathbf{W}}_\theta = \mathrm{diag}(\theta_1, ..., \theta_n)$ with the adaptive parameters θ being now the filter in the spectral domain which is *diagonal* as the filter acts pointwise.

Even though this formulation looks powerful, it has a few limitations. First of all, it is not guaranteed that the learned kernels in the spectral domain $\hat{\mathbf{W}}_\theta$ are localized in space, i.e. act locally in the spatial domain instead of globally. In fact, due to their discrete eigenvalues, they will rather act globally.

Furthermore, the filters are domain-dependent, since filters trained by a particular graph depend on the eigenvectors of $\mathbf{L}$. Therefore, the filters rely on the structure of the graph and will not generalize to graphs with a different structure. Among these limitations, also the number of parameters scales with the number of nodes n.

10.5.1.1 *Stable and localized filters*

To overcome these limitations, $\hat{\mathbf{W}}$ can be approximated using a smooth spectral transfer function $\tau_\theta(\lambda)$ which depends on the eigenvalues, where θ are adaptive parameters [97]. To design a filter less dependent on the eigenvectors of $\mathbf{L}$, a series expansion for $\tau_\theta(\lambda)$ can be chosen. When using polynomials, the approximation reads

$$\tau_\theta(\lambda) \approx \sum_{i=0}^{k-1} \theta_i \lambda^i \tag{10.10}$$

and terminates after some number k, which controls the complexity of the filter.

Applying this approximation, the stable and localized convolution reads:

$$\mathbf{\Phi}\hat{\mathbf{W}}_\theta\mathbf{\Phi}^T f \approx \mathbf{\Phi} \begin{pmatrix} \tau_\theta(\lambda_0) & & \\ & \ddots & \\ & & \tau_\theta(\lambda_{n-1}) \end{pmatrix} \mathbf{\Phi}^T f \tag{10.11}$$

Among the fact that with such a modification, the number of parameters is independent of the number of nodes of a graph, it is important to approximate $\tau_\theta(\lambda)$ as smooth, since a smooth function in the spectral domain corresponds to a localized filter in the spatial domain.

Fast filtering: However, even though this formulation allows for stable and localized filters, it is computationally expensive due to the eigenvector decomposition of $\mathbf{L}$. Thus, for data that feature different graph structures, the eigendecomposition would have to be carried out individually for each graph.

To accelerate the computation, an explicit expansion of the Chebyshev basis can be used [98] instead of expanding the spectral transfer function as series in the spectral domain. By rewriting the convolutional operation

$$\mathbf{\Phi}\hat{\mathbf{W}}_\theta\mathbf{\Phi}^T f = \mathbf{\Phi}\hat{\mathbf{W}}_\theta(\mathbf{\Lambda})\mathbf{\Phi}^T f = \mathbf{W}_\theta(\mathbf{L})f,$$

and using an approximation of $\mathbf{W}_\theta(\mathbf{L})$ instead of $\hat{\mathbf{W}}_\theta(\mathbf{\Lambda})$

$$\mathbf{W}_\theta(\tilde{\mathbf{L}})f \approx \sum_{i=0}^{k-1} \theta_i T_i(\tilde{\mathbf{L}})f, \tag{10.12}$$

the convolutional operation for so-called *ChebNets* can be defined. Here, T_i are the Chebyshev polynomials which form an orthogonal basis in $[-1, 1]$. Hence, the Laplacian is rescaled via $\tilde{\mathbf{L}} = \frac{2}{\lambda_{\max}}\mathbf{L} - \mathbb{1}$ so that its eigenvalues are between -1 and 1.

With the recursive definition of the Chebyshev polynomials:

$$T_{n+1}(\tilde{\mathbf{L}}) = 2\tilde{\mathbf{L}} \cdot T_n(\tilde{\mathbf{L}}) - T_{n-1}(\tilde{\mathbf{L}}), \text{ and } T_0(\tilde{\mathbf{L}}) = \mathbb{1}, \; T_1(\tilde{\mathbf{L}}) = \tilde{\mathbf{L}}$$

the expansion can be determined cost-efficiently, as the Laplacian eigenvectors do not need to be determined [99].

The number of adaptive parameters θ in this convolution depends on k, the order of expansion. This corresponds not only to the complexity of the filter, but additionally, how neighbors at which distance are affected by the convolution. Remember that smoothness in the spectral domain corresponds to locality in the spatial domain, and further, that using powers of the adjacency $\mathbf{A}^k$ includes information over k-hop neighbors. For example, for $k = 0$, only the 'central node' is considered, for $k = 1$ all adjacent neighbors, and for $k = 2$ the neighbors of the neighbors and so on.

Beyond this behavior, the transition from ChebNet motivated by spectral methods to spatial methods can be shown. The propagation rule for GCN [77] (discussed in Sec. 10.4.1) can be derived from a first-order approximation of (10.12) (see Example 10.6). Based on this derivation, each GCN can also be understood as a small transformation acting on the spectrum of the graph Laplacian. Furthermore, Laplacian based methods can be formulated as message passing neural networks (see appendix of Ref. [82]).

Example 10.6. From the spectral to the spatial domain: Although networks based on convolutions acting in the spectral domain seem to be less connected to networks utilizing the convolution in the spatial domain, there is a relationship. Starting from (10.12) using a first-order approximation (evaluating for $k = 2$) results in:

$$\hat{\mathbf{W}}_\theta(\tilde{\mathbf{L}})f \approx \sum_{i=0}^{k-1} \theta_i T_i(\tilde{\mathbf{L}})f$$

$$\approx \theta_0 f + \theta_1 (\mathbf{L} - \mathbb{1})f$$

$$\approx \theta_0 f - \theta_1 \mathbf{D}^{-\frac{1}{2}}\mathbf{A}\mathbf{D}^{-\frac{1}{2}}f,$$

by setting $\lambda_{\max} = 2$ arguing that the network can adapt to the scale during learning, and identifying the normalized graph Laplacian $\mathbf{L} = \mathbb{1} - \hat{\mathbf{D}}^{-\frac{1}{2}}\hat{\mathbf{A}}\hat{\mathbf{D}}^{-\frac{1}{2}}$. If we further choose $\theta_0 = -\theta_1$, we obtain

$$g * x \approx \theta_0(\mathbb{1} + \hat{\mathbf{D}}^{-\frac{1}{2}}\hat{\mathbf{A}}\hat{\mathbf{D}}^{-\frac{1}{2}})f$$

$$\approx \hat{\mathbf{D}}^{-\frac{1}{2}}\mathbf{A}\hat{\mathbf{D}}^{-\frac{1}{2}}f\theta_0,$$

which is similar to the result for the propagation rule obtained for the GCN, which features self-connections of the central node, i.e. $\hat{\mathbf{A}} = \mathbb{1} + \mathbf{A}$. Generalizing the concept to multi-dimensional input channels results in (10.3).

Although ChebNet allows for fast and localized filtering, a limitation of many spectral methods is that due to the symmetric Laplacian, the convolution in the spectral domain corresponds to a convolution in the spatial domain using symmetric kernels only.

Physics applications: For spherical data structured following the Hierarchical Equal Area isoLatitude Pixelization (HEALPix) [100], often used in cosmology, the DeepSphere architecture was proposed [101]. DeepSphere utilizes a graph based on the HEALPix to define via the Chebychev expansion of the rescaled graph Laplacian [98] a convolutional operation on a sphere. For data beyond the HEALPix, SphericalCNNs based on the Fast Fourier Transform (FFT) [102] or architectures utilizing edge convolutions (see Sec. 10.4.2) can be used.

10.6 Exercises

Exercise 10.1. Signal Classification using Dynamic Graph Convolutions: See the description of a cosmic-ray observatory in Example 11.2 and Fig. 11.2(b). After a long journey through the universe before reaching the earth, the cosmic particles interact with the galactic magnetic field $\vec{B}$. As these particles carry a charge q they are deflected in the field by the Lorentz force $\vec{F} = q\,\vec{v} \times \vec{B}$.

Sources of cosmic particles are located all over the sky, thus arrival distributions of the cosmic particles are isotropic in general. However, particles originating from the same source generate on top of the isotropic arrival directions, street-like patterns from galactic magnetic field deflections.

For a Dynamic Graph Convolutional Neural Network start with the script (Exercise 10.1) from the exercise page. The goal is to classify whether a simulated set of 500 arriving cosmic particles contains street-like patterns (signal), or originates from isotropic arrival only (background).

Summary 10.1.

(1) Physics experiments often feature data lying on non-regular or non-Euclidean domains due to complex sensor placement or challenging phase spaces of underlying physics processes.

(2) By interpreting such complex data as graphs, consisting of nodes and edges, operations utilizing convolutions on such data can be realized using graph convolutional neural networks.

(3) Several algorithms exist to utilize convolutional operations on undirected, directed, or weighted (embedded) graphs. Therefore, the adjacency matrix $\mathbf{A}$ or graph Laplacian $\mathbf{L}$ needs to be determined. They are defined by the neighborhood relations of the graph.

(4) Graph convolutional approaches can be separated in spatial methods, which define filtering operations using kernels localized in the spatial domain and spectral methods, which perform the convolution in the Fourier domain.

(5) The graph convolutional network (GCN) allows for a transition from spectral and spatial methods. Its simple structure enables versatile applications on undirected graphs.

(6) For data distributed as point clouds (embedded data), edge convolutions can be utilized to design sophisticated algorithms. Here, the symmetry of the convolutional operation can be defined by the user. The concept of edge convolutions can be extended with dynamic graph convolutional neural networks (DGCNNs). Here, a clustering of neighborhoods beyond the prior embedding input to the network can be exploited, which allows for a convolution that convolves nodes not determined by local proximity but by similarity.

(7) Using spectral graph convolutions, especially data that feature the same graph structure, e.g. measurements of a detector featuring a non-Euclidean but fixed sensor placement, can be analyzed. Therefore, the convolution is acting on the spectrum of the graph Laplacian.

(8) The proposed methods feature, if at all, a small domain dependency and thus can be used for applications with data sets featuring different graph structures.

Chapter 11

Multi-task learning, hybrid architectures, and operational reality

Scope 11.1. In physics data analyses, we often aim to solve various tasks simultaneously. In this chapter we introduce strategies for multi-task learning. We also combine information from multiple devices. To master different input data structures, combinations of various network concepts are frequently employed. Also, the experiments' operational reality is a challenge on its own and can be taken into account during training. Finally, the performance of the networks must be verified under real-world conditions. In this chapter, we present typical considerations for data-driven knowledge discovery using neural networks.

11.1 Multi-task learning

So far, we gave the networks a single task to solve, such as a classification or regression. Often we have other related tasks that we could also solve with the network. Do we build an appropriate network model for each task or a common model for the various questions?

Although the particular question explores specific aspects of the data, this information may help solve another task. By coupling tasks, a network may favor or suppress model hypotheses over alternative hypotheses during training. An example is studying white and colored noise signals while simultaneously analyzing occasional signals above the noise. By knowing the noise precisely, the signal traces may be better separated.

Multi-task learning always occurs when multiple objective functions are used simultaneously. It is often observed that the network models overtrain less and generalize better. Figure 11.1(a) shows the most common form of

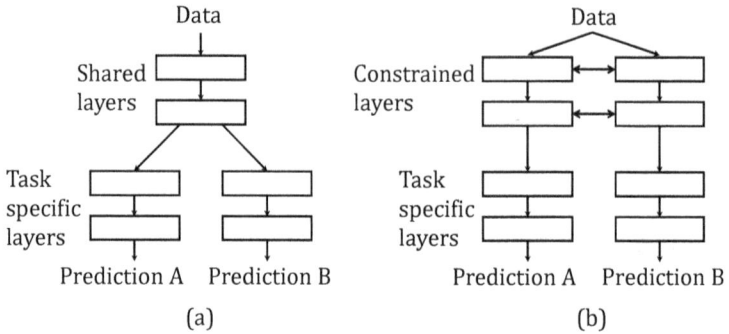

Fig. 11.1. Multi-task learning: (a) hard parameter sharing, (b) soft parameter sharing.

so-called *hard parameter sharing*, where a part of the network is shared across all tasks before subnetworks follow for each task. In so-called *soft parameter sharing* [Fig. 11.1(b)], each task has its own subnetworks, but the parameters are kept similar in size via regularization. Several further developments are reviewed in Ref. [103].

To perform multi-task training, the objective function needs to be modified to combine the objectives of all tasks to be solved. If, for example, a network is trained to solve a task A and a task B, with objective functions $\mathcal{L}_A, \mathcal{L}_B$, the final objective reads

$$\mathcal{L}_{\text{tot}} = \mathcal{L}_A + \lambda \mathcal{L}_B, \qquad (11.1)$$

where the hyperparameter λ is to be chosen by the user.

11.2   Combination of network concepts

Physics experiments often combine different types of information, each with its own specific data structure. It makes sense first to process each of these data types in a subnetwork with suitable architecture and then combine and further process the resulting information.

An example is a video camera with an array of $n \times n$ silicon sensors whose signals $S_i(t)$ are read out as a function of time t. At fixed time t_o, the n^2 signals can be interpreted as an image using a convolutional network. Concurrently, each sensor's signal shape over time may also be of interest and can be analyzed using a recurrent network.

11.3 Operational reality and verification

Many experiments in physics are prototypes, meaning that not everything works exactly as intended. To use neural networks under *real operating conditions*, they must be robust to local biases and partial failures. The effect of operational reality can be built into the training data or should be reflected in the network model.

An important task of us physicists is to quantify the uncertainties of measurements. Indispensable for the reliability of results obtained with networks is, therefore, the verification of the network models over the phase space of the input variables used. This can happen, for example, through independent measurements and their evaluations.

Most often, neural networks are trained using simulated data that do not match the measured data quite exactly. We refer the reader to Chapter 19 on domain adaption and refinement to deal with these differences.

11.4 Physics applications

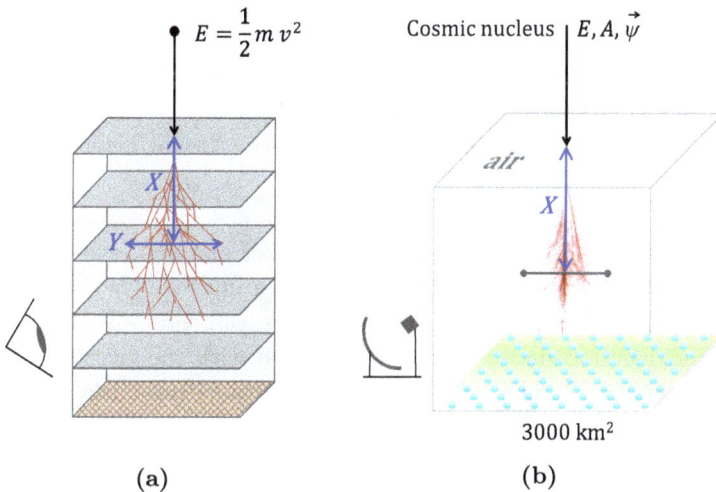

(a)   (b)

Fig. 11.2. (a) Thought experiment for testing new absorber materials. (b) Pierre Auger Observatory in Argentina for measuring air showers induced by cosmic nuclei with energy E, mass number A, and arrival direction $\vec{\psi}$ near normal incidence.

Example 11.1. Test of new absorbing materials: Figure 11.2(a) shows a thought experiment for testing new absorbing materials. Objects with defined energy E fall from a tower onto the material. Relevant is the maximum lateral extent Y and the corresponding penetration depth $X(Y)$. For a transparent material, the impact tests can be documented with a camera. To measure opaque materials, $n \times n$ highly sensitive vibration sensors are mounted on the lower plane. Each sensor measures the local vibrations as a function of time. With all sensors, the scattering of objects until their eventual absorption in one of the layers can be traced, and the penetration depth and lateral extension can be calculated.

Multi-task-learning: The goal of the network is to predict the maximum lateral extent Y and its penetration depth $X(Y)$. Besides, the network shall predict the initial energy E of the object at impact. Simulations of the impact experiment, the absorber material, and the sensors are used to generate appropriate training data and train the network.

Combination of network concepts: First, the time trace of each vibration sensor is analyzed using a recurrent subnetwork and is characterized by ten intermediate variables. These variables are further analyzed in a subsequent fully-connected subnetwork before splitting into two fully-connected subnetworks. One subnetwork is used to determine the object energy, the second to predict the lateral extent Y and its penetration depth $X(Y)$ [compare with Fig. 11.1(a)].

Verification: First, a transparent absorber material is used to obtain independent measurements from the camera images. The predictions of the network are compared with these images. Similarly, the object energy predicted by the network can be compared with the energy calculated by the falling height. After verification, opaque materials can also be measured.

Example 11.2. Signatures of cosmic-ray induced particle showers in the Earth's atmosphere: In this example, we present the Pierre Auger Observatory, located in Argentina in the province of Mendoza that extends over an area of 3,000 km^2 [Fig. 11.2(b)]. Here, air showers induced by cosmic particles (atomic nuclei) colliding with molecules of the Earth's atmosphere are measured. Wanted are the sources of the cosmic particles.

At cosmic particle energies above 10^{18} eV (electron volts), billions of secondary particles are generated and hit the Earth's surface on several square kilometers. Because of the secondary particles' high statistics, comparatively few measuring stations on Earth are sufficient to measure these air showers. The stations are spaced 1.5 km apart in a hexagonal array and each consist of a water reservoir with light sensors. The sensors record the Cherenkov light as a function of time produced by the secondary particles as they pass through the water at speed greater than that of light. A simulated signal trace of a light detector is shown in Fig. 11.3.

On moonless nights, the air showers and their longitudinal evolution are additionally observed with telescopes that pick up ultraviolet fluorescent light from air molecules that are hit.

Multi-task learning: To characterize the cosmic particles, its direction of arrival $\vec{\psi}$, the total energy E in the air shower, and the penetration depth X are determined, where X specifies the height of the maximum particle emission from the air shower. The depth X is corrected for the varying air density as a function of the height above ground and provides information about the mass number A of the cosmic nucleus. Also, the center-of-gravity of the shower impact on the ground, the so-called shower core, can be reconstructed.

All quantities can be measured by Cherenkov light measurements in the water reservoirs [104]. For this purpose, the time of the first signal registered at each station is processed in a network for the angle of arrival. Also, the signal traces of all light sensors are evaluated as a function of time.

Combination of network concepts: First, the signal traces are evaluated sensor-wise using the same recurrent subnetwork (bidirectional LSTM cells, see Chapter 9) and characterized into ten variables. Subsequently, the arrival times at the stations are processed together with the ten variables. Here, the hexagonal symmetry of the station positions is taken into account by a hexagonal convolutional network (see Sec. 8.5).

Finally, the arrival direction $\vec{\psi}$, total energy E, and penetration depth X are reconstructed in three task-specific fully-connected networks [as in Fig. 11.1(a)].

Operational reality: Occasionally, a single station will fail to send a signal due to a defect in the light sensors or the electronics, resulting in a lack of information for the reconstruction of the air shower. Similarly, a light sensor may be saturated if a huge number of secondary particles pass through a water reservoir. As a consequence, a too-small signal is reported by the station. These and other real experimental operation examples can be incorporated directly into the training data, making the network robust to defects and measurement limitations (e.g. by masking stations).

Verification: The telescope system of the observatory operates independently of the Cherenkov-light measurements in the water reservoirs. This allows all network predictions to be evaluated and verified for the sub-datasets of the nightly measurements.

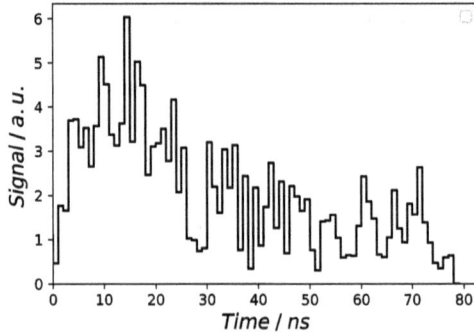

Fig. 11.3.   Signal trace of a light sensor recording Cherenkov light of particles passing a water reservoir.

11.5   Exercises

Exercise 11.1. Cosmic rays: Follow the description of a cosmic-ray observatory in Example 11.2 and Fig. 11.2(b). The simulated data contain 9×9 detector stations which record traversing particles from the cosmic-ray induced air shower. Each station measures the following two quantities, which are stored in the form of a map (2D array) corresponding to the station positions in offset coordinates (Fig. 8.12):

- *time*: time point of detection in seconds
- *signal*: signal strength in arbitrary units

The following shower properties need to be reconstructed:

- *showeraxis*: x, y, z unit vector of the shower arrival direction
- *showercore*: position of the shower core in meters
- *logE*: $\log_{10}(\text{energy} / \text{eV})$

Reconstruct the properties of the arriving cosmic rays by analyzing their air showers:

(1) Set up a multi-task regression network for simultaneously predicting shower direction, shower core position, and energy. The network should consist of a common part to the three properties, followed by an individual subnetwork for each property. Combine the mean squared errors of the different properties using weights λ_i.

(2) Train the model to the following precisions:

- Better than 1.5 deg angular resolution
- Better than 25 m core position resolution
- Better than 10% relative energy uncertainty

Estimate what these requirements mean in terms of mean squared error loss and adjust the relative weights in the objective function accordingly.

(3) Plot and interpret the resulting training curves, both with and without the weights λ_i in the objective function.

Summary 11.1.

(1) The coupling of different tasks in the same network gives priority to aspects in the data during training that would not have been given priority when solving a single task. This so-called *multi-task learning* has a regularizing effect, can lead to less overtraining, and better generalization of the networks. It can be realized by common subnetworks for all tasks (*hard parameter sharing*). Alternatively, parameters can be constrained between the subnetworks for individual tasks (*soft parameter sharing*).

(2) In data analyses in physics, often quite heterogeneous data structures are used for a task, e.g. consisting of time-dependent signals together with integral quantities. To optimally process individual data structures and their symmetries, a single network concept is often not sufficient. Instead, *hybrid architectures* are used in which subnetworks process individual data properties with the appropriate architecture.

(3) Physics experiments are frontier research and, therefore, mostly prototypes. Under real operational conditions, small defects and limitations are registered occasionally. When a network is trained by simulation data that assumes ideal conditions, erroneous network predictions may occur. To prevent this, the networks are made robust for *real operating conditions* by deliberately incorporating randomly occurring defects. If necessary, fine-tuning the simulations to the measured data's properties should also be considered (Chapter 19).

(4) To *verify* the network predictions, it helps to enable independent test scenarios, e.g. by a second measurement setup. This allows quantifying the uncertainties of the network predictions over the relevant phase space of the input variables.

PART 3

Introspection, Uncertainties, Objectives

In this part, we present methods to gain insight into the working principles of neural networks and to interpret their predictions. Furthermore, we explain the quantification of uncertainties that arise when analyzing data with deep networks. Finally, we discuss objective functions from the probability perspective and criteria for their selection.

Chapter 12

Interpretability

Scope 12.1. To apply neural networks in a justified way, we need to understand how predictions of a neural network are produced. This requires techniques that allow us to look into the networks. First, we present two approaches that provide insights into the hierarchical order of information processing by visualizing learned properties within the network. Second, we discuss strategies that allow us to interpret network predictions. Here, we first present a method addressing the input data's most decisive influence on a network prediction. The second type of strategy assigns so-called attributions to the input data that quantify their impact on the network prediction.

12.1 Interpretability and deep neural networks

Since the beginning of deep learning there has been, along with the desire to build more powerful models, also a growing interest in obtaining an in-depth understanding of these algorithms. The resulting collection of techniques enabling the interpretation of machine learning algorithms is large and versatile. In this chapter, we give an overview of relevant methods and introduce the general concept of network introspection. Therefore, we provide the reader with a basic understanding of model interpretability and introduce techniques for the introspection of neural networks. For a comprehensive overview, refer to Ref. [105, 106].

In principle, the interpretability of machine learning models can be divided into three different categories of understanding:

- Data: *Which input is most useful, which part of the data is important?*
- Model: *How is the model working? What is learned by a particular layer?*
- Predictions: *Why is a certain class/value predicted, and what contributes how much to this prediction?*

Unfortunately, it is challenging to examine the individual aspects in practice, as they are strongly interconnected.

Using data, we can introspect network components and gain insight into how information is processed in the network. This is addressed in the first part. Here, we obtain information about the model and its interpretability by visualizing features of individual layers, channels, and nodes. Through this, we gain an understanding of why neural networks are so powerful despite their comparatively simple components and structures.

We then ask what aspects of the input data led to the network predictions. In doing so, we analyze, on the one hand, the sensitivity of the network to changes in the data. On the other hand, we assign to the input data so-called attributions indicating their quantitative influence on the network prediction. In this way, we investigate the interpretability of predictions. This is particularly relevant for physics research, as understanding the predictions is crucial for building a trustworthy model.

To provide a visual understanding, we use convolutional networks in the following. Still, most of the methods can be applied to various network architectures.

12.2 Model interpretability and feature visualization

To understand the functioning of deep neural networks, it makes sense to decompose the network into smaller building blocks first and investigate the working principle of single operations before studying their interaction. In principle, a visualization of the weight matrix and the resulting activations could be performed. Still, beyond the first layer, weights and activations are hard to understand by humans. This limit is explored in-depth in Exercise 12.1. Thus, for CNNs (see Chapter 8), it is obvious to study the working principle of single neurons, channels, and layers. For the upcoming discussion, remember that CNNs work with filters that deposit their results in feature maps (see Fig. 8.1). Two methods that enable human-interpretable visualization of the individual network components are discussed below.

12.2.1 *Deconvolutional network*

One of the first approaches for visualizing deep features (features beyond the first layer) in CNNs, which utilize the ReLU activation function [Fig. 5.8(a)], was proposed in Ref. [107] using a *deconvolutional network* (DeconvNet). The basic idea is to map a single activation of an arbitrary feature map back to the input space to visualize which pattern caused the observed activation. Therefore, we attempt to 'invert'[1] the flow of CNNs.

To find a distinct pattern to which a specific feature is sensitive, samples from a validation data set are chosen that induce the highest feature activation. This allows for a first assessment of its respective characteristic by understanding what maximizes (or minimizes[2]) a specific feature. Further, finding an exceptionally high activation in a feature map leads us to assume that the respective data sample gives a strong response in the filter that was used to analyze the previous layer.

All other feature maps in the selected layer are set to zero to map the activation back to the input space. In the selected feature map, all activations but the highest one are zero-filled. Then, the single activation is mapped backward to produce a pattern in the input space that caused the particular activation. As the signal is propagated all the way to the input, we can represent it as a natural image. The approach is illustrated in Fig. 12.1.

To invert the data flow in CNNs, the following three operations are used:

Transposed convolutions: Using transposed convolutions, the feedforward pass of the convolutional operation is reversed (see Sec. 8.4), allowing a mapping from the feature space to the input space. Therefore, the weight matrices of the already trained model are used but flipped horizontally and vertically.

Activations and zero-clipping: To account for the ReLU activation function in the model, after each transposed convolution, all negative values in the obtained feature maps are set to zero.

---

[1]Because of non-bijective operations, e.g. activation functions like ReLUs, or pooling layers, an inversion can only be approximated.

[2]Minimization of feature responses is also a method for understanding deep networks, but since negative activations are usually cut away by activations (ReLU), maximization is used more often.

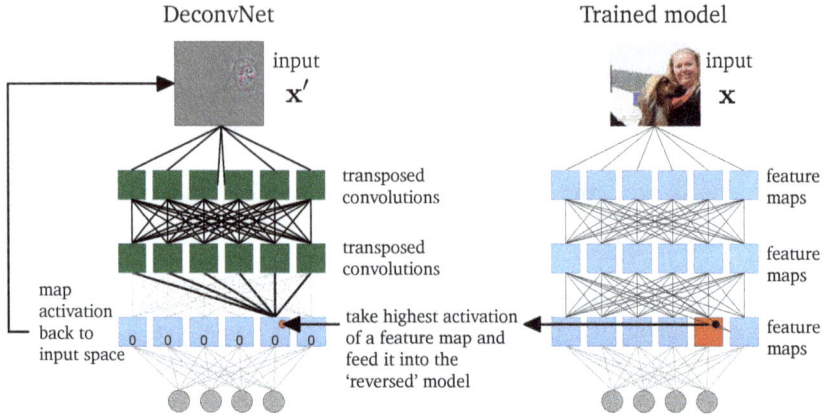

Fig. 12.1. Principle of the DeconvNet approach. A single activation in a feature map, is traced back through the model causing a pattern in the input that corresponds to the induced activation. Data from Ref. [107].

Switches: So-called switches are used to 'invert' max-pooling operations. In the forward pass, the positions of the highest activations are stored in the switches permitting to restore the positions during the 'deconvolution process'. All other activations are set to zero.

Using this technique, features at different layers in the network could be visualized for the first time, yielding a visual impression of the hierarchy learning concept (Fig. 12.2). Here, in the first layers, semantically global but spatially local low-level representations are learned, e.g. filters for edge detection, Gabor filters (orientation detection), and filters for colors. In later layers, the features become more abstract and exhibit a less global meaning. In the final layers, the semantic meaning of different filters is specific and strongly class-related.

Interestingly, using DeconvNet, the winner of the ILSVRC'12[3] image classification challenge, the so-called AlexNet [37] was visualized, and weak spots of the network were identified. Submitting a refined version with changed filter sizes and a new normalization led to winning the challenge in 2013.

---

[3]ImageNet Large Scale Visual Recognition Challenge.

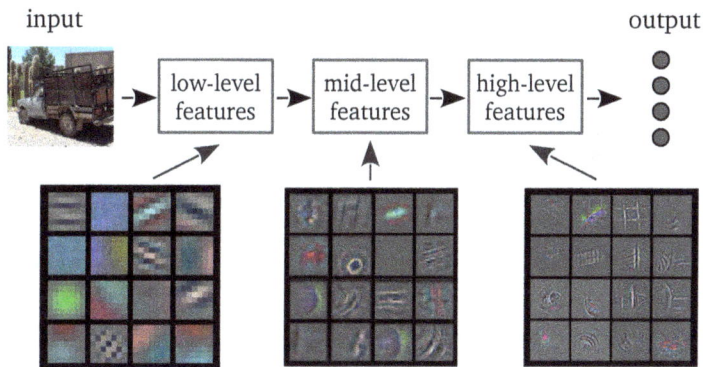

Fig. 12.2. Visualization of features using DeconvNet. The visualized patterns correspond to visualized features of low and deep layers. Data from Ref. [107].

12.2.2 *Activation maximization*

The core idea of *activation maximization* is to find an input pattern that maximizes a specific feature using the differentiability of the model. Compared to the DeconvNet approach discussed above, this allows for a more general procedure as we do not need to search for samples that produce high activations in the validation data set.

The overall target for finding an input image $\tilde{\mathbf{x}}$ which maximally stimulates a single activation can be formulated as:

$$\tilde{\mathbf{x}} = \arg\max_{\mathbf{x}} h^l_{i,c}(\mathbf{x}, \theta) \tag{12.1}$$

Here $h^l_{i,c}$ denotes the activation at pixel location i, channel c, and layer l, further assuming a pre-trained network f_θ holding weights θ which are fixed. When starting with a random input $\mathbf{x}_0$, consisting of Gaussian noise, the gradient with respect to each pixel in the input $\mathbf{x}$ can be obtained using:

$$\mathbf{g} = \left. \frac{\partial f(\mathbf{x}, \theta)}{\partial \mathbf{x}} \right|_{\mathbf{x}=\mathbf{x}_0} \tag{12.2}$$

As we estimate the gradient with respect to the model input $\mathbf{x}$, the resulting gradient $\mathbf{g}$ can be presented in the same way as the input data (an RGB image in this case).

Since we want to find an input that *maximizes* the stimulation, we can make use of simple *gradient ascent*, by adding to the starting image the obtained gradient scaled by a parameter α with a small value[4]

$$\mathbf{x_0}' = \mathbf{x_0} + \alpha \left. \frac{\partial f(\mathbf{x}, \theta)}{\partial \mathbf{x}} \right|_{\mathbf{x}=\mathbf{x_0}}. \qquad (12.3)$$

Repeating this procedure iteratively until convergence results in an input, which maximizes the respective activation. An example visualizing the sensitivity of a neuron is shown on the left in Fig. 12.3.

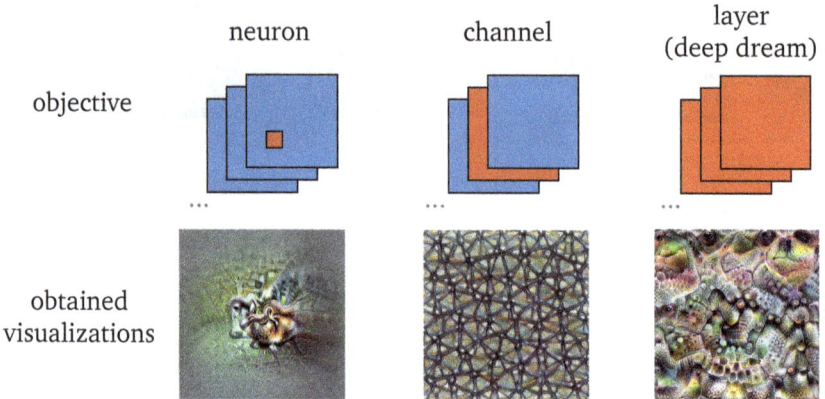

Fig. 12.3. Different objectives for feature visualization. Samples are obtained with activation maximization and the InceptionV1 architecture. Reproduced using [108].

Visualization objectives and deep dream: Revisiting Eq. (12.1) shows that various other visualization possibilities can be investigated. As already discussed, for a single activation (neuron) an input can be obtained which excites this respective neuron. Additionally, a channel-wise feature visualization can be found that excites a full channel of a feature map using the objective function $\mathcal{L}_c^l = \sum_i h_{i,c}^l(\mathbf{x}, \theta)$. To even study the working principle of a basic building block of deep neural networks — a neural network layer — the *deep dream* objective function can be utilized [109, 110]:

$$\mathcal{L}_{\text{dream}}^l = \sum_{i,c} \left(h_{i,c}^l(\mathbf{x}, \theta) \right)^2 \qquad (12.4)$$

With this layer-wise objective function, complex features can be visualized. These methods have gained popularity for a wider range of applications as the results remind of faces, animals, and clouds.

---

[4]This is similar to the learning rate introduced in Sec. 4.7.

In Fig. 12.3, the objectives are compared, and examples are shown for neuron-, channel-, and layer-wise visualizations. Note that the CNN filters are not visualized themselves. Instead, we observe to which patterns the filters (neurons, channels, layers) are sensitive. The examples were obtained using the InceptionV1 architecture (see Sec. 8.4) and the lucid software [108]. It can be seen that some of the visualizations feature structures, which resemble natural images with which the InceptionV1 network was trained, e.g. a dog face in the right picture is visible. However, the visualized features show rather complex structures, which are far beyond Gabor filters. In particular, the sensitivity of the layer looks exceptionally complex and versatile.

In principle, the features of the model's final layer,[5] can be used for model visualizations as well. With such an approach, starting with random noise, patterns can be generated which typically belong to a particular class.

Currently, many different approaches exist to visualize extracted features of deep neural networks, e.g. by finding negative neurons, studying feature diversity, or their complex interplay, etc. [111–114]. We also encourage the reader to explore the in-depth overviews in Ref. [115–117].

12.3   Interpretability of predictions

Next, we consider methods to interpret predictions. We start with a trained model f_θ used to give a prediction $f_\theta(\mathbf{x_0})$ for a given sample $\mathbf{x_0}$. Now, we want to study what causes the network to form exactly this prediction. Because such an approach investigates the behavior of f_θ locally around the respective sample $\mathbf{x_0}$, such techniques are often referred to as *local methods*.

In the following, we first introduce simple methods to study the sensitivity of a model, e.g. by investigating to which input the network is particularly sensitive. In the second part, we consider more sophisticated methods which rely on attribution, and thus, permit us to quantify the contribution of each input to the final prediction. Finally, we briefly investigate methods that rely on perturbations and can be utilized beyond neural networks. For simplicity, we focus on the interpretation of classifier predictions, but most methods can also be used for regression models.

---

[5]Due to numerical stability, usually the layer with pre-activations (before applying the softmax) is utilized.

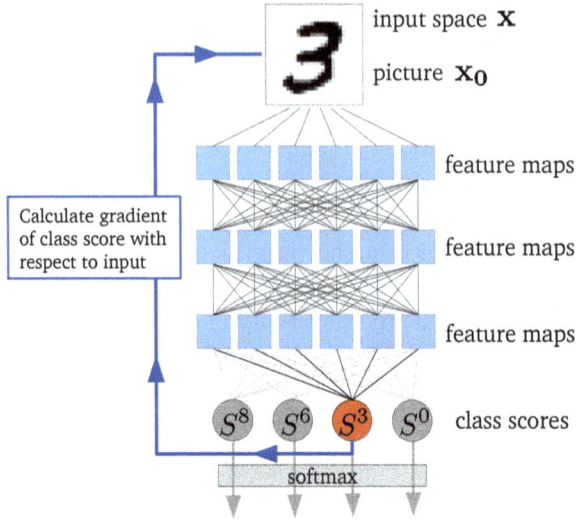

Fig. 12.4. Working principle of saliency maps illustrated for a network trained on the MNIST data set [118].

12.3.1 *Sensitivity analyses and saliency maps*

A simple approach, which does not require a specific architecture, consists of building *saliency maps*. Here, the differentiability of neural networks is explicitly exploited to enable post-hoc interpretations of predictions. The idea is to interpret the model's prediction[6] by studying the gradient **g** with respect to the input to assess which information (e.g. which pixels) has a significant influence on the prediction.

Now we interpret the prediction $S^c(\mathbf{x_0}, \theta)$ of the model for the image-like sample $\mathbf{x_0}$. Here, $S^c(\mathbf{x_0}, \theta)$ denotes the predicted *class score* (i.e. the classification output before applying a softmax activation) of f_θ for a class c. For each component of the input **x** (e.g. image pixels), the gradient can be obtained using:

$$\mathbf{g}^c = \left.\frac{\partial S^c(\mathbf{x}, \theta)}{\partial \mathbf{x}}\right|_{\mathbf{x}=\mathbf{x_0}} \tag{12.5}$$

That means, we estimate the derivative of the class score with respect to the input image and evaluate it at point $\mathbf{x_0}$, the sample whose prediction we want to analyze. Hence, saliency maps explain a prediction by quantifying

---

[6]By extending the application beyond the class score, it is also possible to use saliency maps for visualization of features by examining individual activations.

how it varies with the modification of the input. A visualization of the technique is shown in Fig. 12.4.

Since $\mathbf{g}^c$ has the same format as the model input, the obtained gradients can be presented as the input, e.g. as an image. This visualization of the gradient is called a *saliency map* or *sensitivity map* as it visualizes to which input the network is particularly sensitive. Since a saliency map can be created for each class, sensitivity studies can be performed class-wise.

The larger the observed absolute gradient $|\mathbf{g}^c|$, the stronger the sensitivity of the model to the respective pixel of $\mathbf{x}_0$. Besides, the sign of the gradient is relevant. Increasing the intensity of a pixel with a positive gradient will increase the prediction, whereas an increase of a pixel with a negative gradient will make the prediction decrease. We address this difference in Example 12.1.

Example 12.1. Saliency maps and handwritten digits: Let us consider a network trained on the MNIST data set (which consists of handwritten digits), and thus, can be easily interpreted by humans. Figure 12.5 shows four examples in the upper row.

In the bottom row, the corresponding saliency maps (for the true classes) are depicted. Although the maps are noisy, they can be used for simple interpretations. Regions with zero gradients are marked green. Yellow and red indicate positive gradients and thus show pixels that have contributed positively to the classification. Look, e.g. at the rightmost figure. The cross in the middle of the picture showing the '8' has a strong influence. This makes sense since the pattern is decisive to distinguish it from a '0'.

Furthermore, we also observe negative gradients, shown in blue. One good example is the salience map of digit 3, where large negative gradients are shown in the center-left part. This could be interpreted as follows: Increasing the pixel intensity here would close the gap and create a pattern similar to a '9'. The interpretation may be carefully paraphrased to: since the image has only low pixel intensities here, the digit to predict is a '3' and not a '9'. Note, however, that saliency maps provide an interpretation of how the prediction changes rather than how it is formed.

input

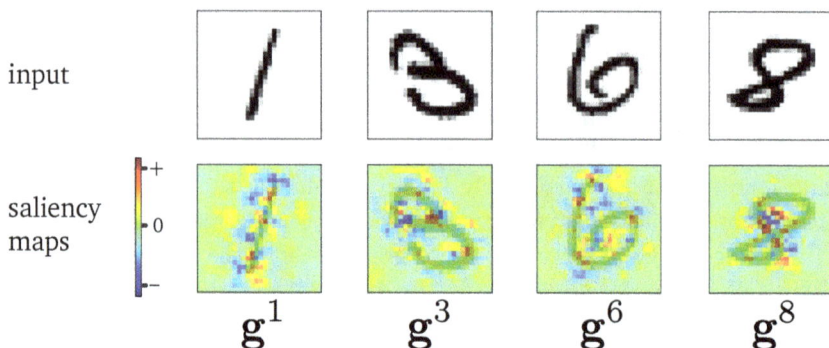

saliency
maps

$\mathbf{g}^1$ $\mathbf{g}^3$ $\mathbf{g}^6$ $\mathbf{g}^8$

Fig. 12.5. Prediction analysis of a network trained on MNIST [118] using saliency maps. Example images (top row) are shown together with the resulting saliency maps (bottom row) which are overlayed with the input.

12.3.2 *Attribution-based prediction analyses*

We further investigate the influence the input data has on the network's predictions. The above-discussed saliency method is based on locally estimated gradients. It is easy to implement and enables reasonable sensitivity studies. However, its absolute expressiveness is limited[7] since it explains *how the prediction would change*, but it does *not explain the prediction itself*.

To enable a more thorough interpretation, recently proposed methods aim to explain the *attribution*, sometimes also called *relevance*, of each input (e.g. image pixel) on the prediction. The formed attributions of the input are called *attributions maps* and can be estimated for each class. In addition to an interpretation of predictions, these maps can be used for semantic segmentation.

When estimating for a given sample $\mathbf{x_0}$ the attribution R_i^c for each input i and specific class c, attribution-based methods fulfill the property that the attributions sum up to the prediction $S^c(\mathbf{x_0})$ of the network, i.e. $\sum_i R_i^c = S^c(\mathbf{x_0})$ [119]. This property is often called *completeness*. Assume, e.g. a classifier that assigns to a galaxy image the class *spiral galaxy* with a probability of 85% and we study the attribution using a method that fulfills the completeness criterion. Then, the attribution of each image pixel is given in [%]. Summing over all pixels yields 85%.

In the following, we present two methods that re-distribute predictions

---

[7]Also, locally estimated gradients tend to be noisy, as visible in Example 12.1.

back to the input: discriminative localization, which is limited to convolutional neural networks, and layer-wise relevance propagation.

Beyond this, a large and growing number of strategies exists. Therefore, we recommend the reader to choose a suitable method for their model and test various techniques and study their results to gain a more extensive understanding of its predictions. Suitable software exists, such as the Python packages *iNNvestigate* [120] or *DeepExplain* [121] that provide various interpretability techniques to introspect networks implemented in Keras. PyTorch users may consider the *Captum* [122] library.

12.3.2.1 *Discriminative localization*

One way to validate predictions is to investigate the produced activations of convolutional layers. The aim is to identify regions that are important for predicting a certain class and are, therefore, decisive for the prediction. For convolutional neural network architectures that feature before the final classification layer a global average-pooling layer, *discriminative localization* [123] can be used.[8]

Remember that using global average-pooling, complete feature maps are aggregated to single nodes. Thus, the output of the model's last convolutional layer reads $h_{i,c}$, where i denotes the pixel position in the map, and c the channel (feature map). Applying global average-pooling leads to a vector $\vec{u}$ whose length is equal to the number of channels n_c. Its components u_c describe the average feature map activation:

$$u_c = \frac{1}{n_{\text{pix}}} \sum_{i}^{n_{\text{pix}}} h_{i,c}. \tag{12.6}$$

Here, n_{pix} corresponds to the number of activations in the feature map. For the particular architectures discussed here, the vector $\vec{u}$ is only scaled with one linear transformation before being fed into the softmax function. Thus, the network output $\vec{y}_{\text{class}}$ reads:

$$\vec{s}^{\,\text{class}} = \mathbf{W}\vec{u} + \vec{b} \tag{12.7}$$

$$\Rightarrow \vec{y}^{\,\text{class}} = \text{softmax}(\vec{s}^{\,\text{class}}), \tag{12.8}$$

where $\vec{s}^{\,\text{class}}$ denotes the vector of class scores.

To create activation maps for each class, the global average-pooling operation is omitted. Instead, the feature maps of the final convolutional layer

---

[8]No additional fully-connected layer may exist in between the pooling operation and the output for this method to work.

are scaled by their respective weights, which indicate their average influence on the model prediction. The multiplication is performed in a pixel-wise manner using the weights $\mathbf{W} = W_{\text{class},c}$. Finally, the bias is added, and the resulting class maps are upscaled to the size of the input image, e.g. using bilinear interpolation (to 'reverse' pooling operations). This yields so-called class activation maps (CAMs). To obtain a rough attribution map,[9] a normalization taking into account the softmax activation function (for a classification task) has to be performed.

This operation is illustrated in Fig. 12.6 for the model trained in Exercise 8.1. Here, the task was to classify the magnetic phase of the system (ferromagnetic or paramagnetic) using a 2D image of spin polarizations. Only the activation map for the ferromagnetic class is shown. We will investigate this application further in Exercise 12.3.

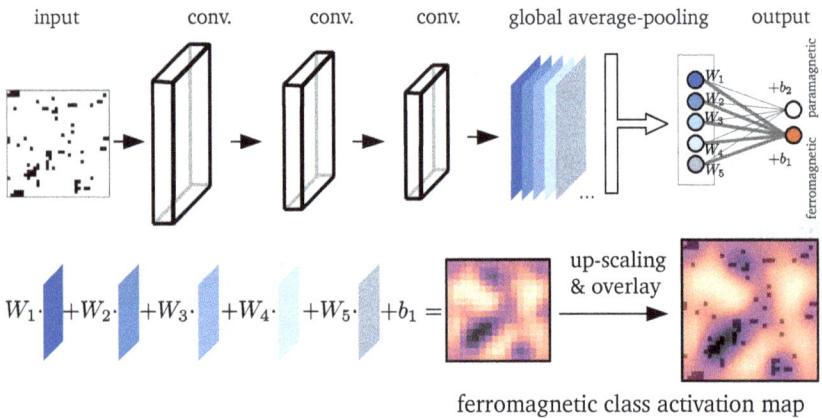

Fig. 12.6. Discriminative localization applied to the classification task of magnetic phases (see Exercise 8.1). For creating class activations maps, the global average-pooling operation is omitted, and the weights of the final layer scale the full activation maps of the last convolutional layer.

Besides classification tasks, the method can also be adapted to regressions and tasks featuring other objectives. Nevertheless, the technique is limited to particular architectures that feature a global average-pooling operation before the output layer.

---

[9]That satisfies the completeness criterion.

12.3.2.2 *Layer-wise relevance propagation*

An alternative strategy suitable for a wider class of network architectures is layer-wise relevance propagation (LRP) [124, 125]. This method can be used, e.g. for fully-connected, convolutional, and even special recurrent networks (LSTMs). Here, the network prediction S^c is redistributed up to the input layer to obtain a *relevance* R_i for each input i. Hereby, it is ensured that the signal that is received by a neuron in the forward pass is fully redistributed in the backward propagation of LRP. This can be interpreted as an analog to Kirchhoff's circuit laws and ensures completeness. Various propagation rules for redistributing relevances exist. One rule, the so-called LRP-ϵ rule, is discussed in Example 12.2.

Example 12.2. LRP-ϵ: The ϵ-rule was derived for networks with ReLU activations. By starting at a prediction, i.e. the relevance at a particular output node (relevances of other output nodes are set to 0), the attribution of each input is calculated as follows. For a layer l, with J neurons j, and relevances R_j^l, the relevance is propagated to the upper layer (towards the input) that holds I neurons i. Now for each node, its particular relevance R_i^{l-1} is estimated. This principle is shown in Fig. 12.7. Since we use ReLU activations (we only have to consider vanishing contribution and linear contributions) the redistribution role reads:

$$R_i^{(l-1)} = \sum_j \frac{a_i W_{ij}}{\epsilon + (b_j + \sum_{i'} a_{i'} W_{i'j})} R_j^{(l)} \tag{12.9}$$

Here, W_{ij} denotes the respective weight to connect a neuron of the upper and the lower layer, b_j the bias, and a_i (and a_i') the activations in layer $l-1$ determined in the forward pass. Iterative application of the propagation rule will lead to the final attribution map **R** at the input that explains a certain prediction.

Whereas the numerator propagates in the forward pass received input of a node backwards, the denominator scales it with respect to all received inputs to conserve the relevance at a given node. The hyperparameter ϵ prevents a division by zero. Choosing large ϵ will absorb weak relevances making the interpretation limited to the strongest influences but gives sparser and less noisy results.

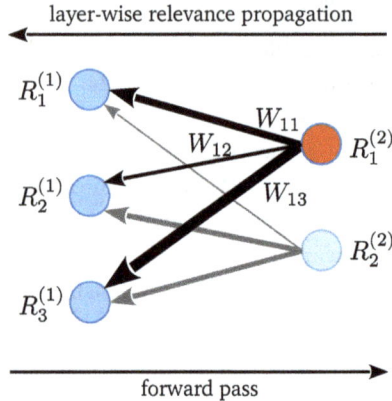

Fig. 12.7. Principle of layer-wise relevance propagation. Activations received in the forward pass are redistributed as relevances R to obtain the attribution of neuron for a particular prediction.

The most meaningful results can be achieved by combining various propagation rules, as empirical studies show that different propagation rules perform better at higher or lower layers. An overview of diverse propagation rules and their combination is given in Ref. [126].

12.3.3 *Perturbation-based methods*

In addition to attribution-based methods discussed above, we can also analyze predictions using perturbation-based techniques [107, 127, 128]. Here, by small modifications of the input, e.g. using patches to occlude parts of the inputs, the influence on the output is studied. By perturbing various parts of the input, smooth attribution maps can be created; however, the computational costs can be tremendous.

Note that when using a simple occlusion method, e.g. as proposed in Ref. [107], on the one hand, the size of the patch influences the results and therefore needs to be chosen carefully.[10] On the other hand, a meaningful replacement has to be chosen (similar to a baseline). For natural images, zeroes (black images) are usually used, which could be, however, a suboptimal choice for a physics data set.

---

[10]This choice depends on the particular data set and the size of its typical patterns.

12.4 Exercises

Exercise 12.1. Weights and activations of a convolutional network: This task consists of two parts. First, set up and evaluate a convolutional neural network.

(1) Starting from the program (Exercise 12.1 — Convolution) from the exercise page, set up and train a convolutional network on the classification of images (data set CIFAR-10).

    (a) Train your network to at least 70% test accuracy.

(2) Plot the confusion matrix. What do you observe?
(3) Plot several falsely classified images along with the predicted class scores. What kinds of misclassification do you observe, why do they occur?

Second, visualize the weights and activations in the network layers. Start from the program (Exercise 12.1 — ConvVisualization) to visualize the inside of the network:

(1) Plot the filter weights in the first layer of your network and see if you can make any sense of it.
(2) Visualize the activations in the first two layers of your network for two input images of choice and describe what you see. Does it meet your expectations?

Exercise 12.2. Visualization by activation maximization: Use the concept of activation maximization (Fig. 12.3) on the MNIST data set of handwritten numbers. Starting from the provided code (Exercise 12.2) perform the following tasks:

(1) Set up a convolutional network (at least three convolutional layers) on the MNIST classification task and train it to at least 98% test accuracy. Provide validation and training curves.
(2) Visualize the feature maps in the convolutional layers using activation maximization. Use Gaussian noise as input and maximize the mean activation of the respective feature maps using 50 steps gradient ascent and a step-size of 1.
(3) Provide the plots of the feature maps and describe/comment on what patterns you see in which layer.

Exercise 12.3. Discriminative localization:

(1) Pick out two correctly and two wrongly classified images classified with a convolutional network.

(2) Look at Exercise 8.1 to extract weights and feature maps from the trained network model.

(3) Create and plot the class activation maps and compare them with the images in order to see which regions lead to the prediction.

Summary 12.1.

(1) Interpretability of machine-learning models involves data, model, and predictions.

(2) In model interpretability, techniques aim to improve the understanding of the functioning of neural networks and the interplay of individual layers and neurons. Here, the behavior of single neurons, features, or layers is studied, and, further, their interplay is evaluated. To visualize the sensitivity of features learned by the model:

- Activation maximization can be used to create inputs that induce a maximal activation, enabling a more *global* understanding of the model.
- *Local* introspection methods like DeconvNet propagate observed signals, which cause strong activations back to the input. Saliency maps utilize gradients to study the model behavior *locally* around a single example and identify inputs that need to change least to change the feature at most.

(3) Simple sensitivity analyses of predictions can be performed by applying local introspection techniques to the model output.

(4) More sophisticated approaches, in particular, estimate the attribution of each input feature to the final prediction. For attribution-based methods, the sum over all attributions is equal to the prediction.

- Using discriminative localization, class activation maps can be created by utilizing the activations in the last convolution layer, estimate their contribution by a proper normalization, and using interpolation to resize the image. This technique is limited to models which feature a global average-pooling layer.
- Layer-wise relevance propagation redistributes observed predictions back to the input and can be used for a variety of networks, including fully-connected, convolutional and recurrent architectures.

(5) Perturbation-based methods estimate the relevance of each input by modifying it and measuring the caused change at the output.

Choosing a proper baseline is essential. Since the techniques usually do not rely on a specific architecture of the model, they can be applied to various machine learning models. However, they can be computationally demanding.

(6) To gain the most comprehensive understanding of a prediction, it is recommended to employ various methods and compare the findings.

(7) When analyzing models to explain them, one always has to be careful to disentangle properties of the algorithms and implications from our human interpretation.

Chapter 13

Uncertainties and robustness

Scope 13.1. Scientific measurements always consist of a measured value and an associated degree of credibility. For example, the best-known value of the inverse fine-structure constant is $\alpha^{-1} = 137.035999084 \pm 0.000000021$ — corresponding to a precision of 0.15 parts per billion. In this chapter, we discuss how traditional uncertainties can be accommodated and potentially mitigated by machine learning and which new uncertainties might become relevant.

While we have an intuition — honed by many undergrad lab courses — of what we mean by uncertainty, we need to introduce some terminology to precisely articulate what is meant in this context. The machine learning literature typically defines two types of uncertainties: *aleatoric uncertainties* and *epistemic uncertainties* [129]. Aleatoric[1] uncertainties are statistical and describe the observed distribution of an experiment repeated with identical initial conditions, such as repeatedly rolling the same dice. Of course, practically, the conditions can never be *exactly* reproduced, leading to the observed uncertainty. Epistemic uncertainties, on the other hand, stem from the incompleteness of our understanding. These are systematic mistakes, such as using the ideal gas law at very low temperatures. In practice, both types of uncertainty are usually present.

Applied to our deep neural networks [130], we refer to noise or scattering inherent in the data as aleatoric uncertainty and the uncertainty of the network due to insufficient training data as the epistemic uncertainty. Accordingly, we expect the epistemic component to decrease with increased

---

[1]From *alea*, Latin for dice.

training data and the aleatoric component to remain approximately constant.

When applying machine learning to realistic tasks in physics, often an additional challenge arises: We might train models on simulations to have greater control over their behavior and to access infrequent scenarios. After training is completed, such a model is evaluated on data recorded by a real detector.

However, simulations are usually not perfect, leading to systematic differences between the training data and the data to which the model is eventually applied (*evaluation data*) which in turn will bias the resulting network predictions. In general, solving the difficulty of a mismatch between training and evaluation data is known as *domain adaptation* and this specific difference is referred to as a *simulation gap*. We discuss solutions to this challenge in Chapter 19.

Another aspect covered here is the *robustness* of network predictions. It becomes relevant when there exists a number of possible perturbations under which we want our output to be invariant. For example, a self-driving car needs to be able to identify correctly a stop sign, independent of the weather conditions. Thus, we often want to reduce the impact of known or unknown perturbations on the output of a trained network. A closely related problem is removing the correlation of network outputs with other observables, for example, if these are known to be poorly modeled.

Finally, we will also study how networks react to inputs specifically crafted to induce a wrong response and how difficult (or easy) it is to construct such *adversarial examples*.

13.1   Measuring uncertainties

Once trained, a simple neural network used for classification usually is a deterministic function of its inputs — assuming, e.g. that no dropout layer is used in the evaluation phase. Therefore, in a sense, no additional uncertainty is added by the network. Unfortunately, this does not mean that we can completely ignore uncertainties when training and using neural networks. We will distinguish between the situation in fields like particle physics, where training often can be done using simulations, and other areas.

If available, machine learning models can be trained on simulations of expected processes and alternate hypotheses. Algorithms are then evaluated on a test dataset to develop a statistical model. Finally, results on the test data — which often consists of statistically independent examples from the same simulation used for training — are compared to recorded experimental data.

Following Ref. [131], we divide the possible uncertainties into:

(1) uncertainties that affect the performance of the network (how well the underlying truth is learned) and
(2) uncertainties that affect how well the prediction from simulation agrees with data (termed bias in this context).

This taxonomy leaves us with four possible combinations: aleatoric/performance, aleatoric/bias, epistemic/performance, and epistemic/bias.

The effects of limited training statistics (epistemic/performance) can be assessed using ensemble methods such as dropout, cross-validation, Bayesian networks,[2] or by performing multiple training runs of a network with different initializations on the same training data. Epistemic effects like a too low network complexity, a wrong training procedure, or a domain shift between training and testing data are more difficult to probe. One approach is to apply small perturbations to the setup and evaluate their impact, for example, by varying different hyperparameters. The above aspects are all uncertainties from the machine learning model's perspective, but not from the view of a statistical analysis using this model as part of its decision-making process.

On the other hand, limited prediction statistics is an epistemic uncertainty affecting the bias of the prediction and can be evaluated using standard statistical methods not specific to deep learning. A default approach to assess this effect is random sampling with replacement — usually termed bootstrapping statistics [132].

Another effect to consider comes from systematic differences between simulated events used for training and real events. These usually stem from the remaining inaccuracies of the simulation tools and are referred to as domain shifts. A standard method to model the impact of a potential domain shift is to vary the inputs within their known uncertainties and take the resulting variation in model output as corresponding uncertainty. We discuss methods to reduce such differences in Chapter 19.

If no precise and well-understood simulations and uncertainty models are available, more general insights from the machine learning literature

---

[2]The name *Bayesian* stems from the explicit use of Bayesian inference in the construction of this method.

can be useful [133]: In general, quantifying epistemic uncertainty is about finding out which regions of input space are not covered by training data (i.e. how much interpolation/extrapolation is needed for a given prediction). Aleatoric uncertainty, on the other hand, is learning about the conditional distribution of a target variable given values of an input variable (i.e. knowing the inherent spread of measurements at a given value) [133].

Again, both cases can be captured by ensemble approaches. Ensembling means that several identical networks are trained for the same problem and their outputs are combined. For uncertainty quantification, we can use two output neurons per network: the quantity of interest (the prediction of a physical observable) and its associated aleatoric uncertainty. The *real* prediction of this ensemble is then the mean predicted value, the aleatoric uncertainty is the mean of the aleatoric output, and the epistemic uncertainty is the standard deviation of the predicted values. The total predictive uncertainty is the squared sum of epistemic and aleatoric components.

Instead of manually training an ensemble of networks and relying on the randomness of weight initialization and stochastic gradient descent, more principled approaches can be used. A promising direction is Bayesian neural networks that replace the individual weight values with probability distributions (see Example 13.1).

Example 13.1. Bayesian neural networks [135] follow a simple idea: Instead of considering each weight as a number optimized during training, we can describe weights with probability distributions. For example, we replace each weight w_i with a Gaussian, parameterized by its mean μ and standard deviation σ:

$$w_i \to \mathcal{N}(\mu_i, \sigma_i^2). \tag{13.1}$$

Practically, this only doubles the number of trainable parameters and turns predictions of the network into probability distributions from which we can sample during training or inference. Figure 13.1 illustrates the idea. Finally, the objective function used for training a Bayesian network needs to be adapted to take into account the usual likelihood minimization and the difference between learned per-weight distributions and their respective priors.

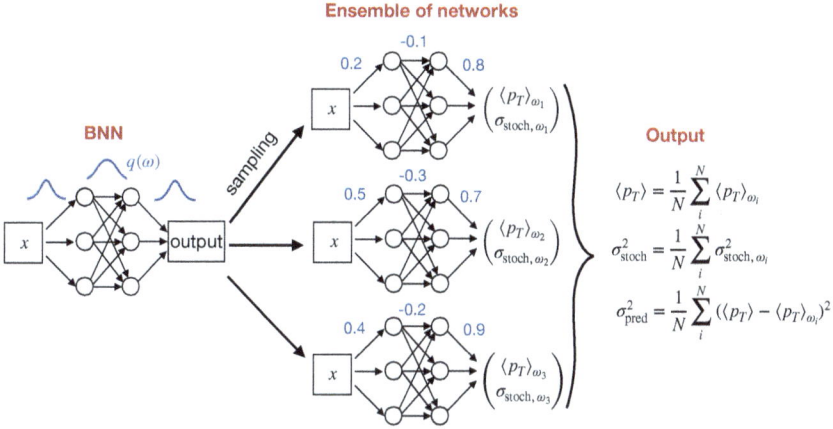

Fig. 13.1. Illustration of a Bayesian neural network for the prediction of a physical quantity (transverse momentum or p_T) and its associated uncertainty from Ref. [134]. Here σ_{stoch} denotes the stochastic or aleatoric uncertainty and σ_{pred} the predictive or epistemic component.

13.2 Decorrelation

Generally speaking, machine learning classifiers achieve higher performance on complex tasks than classical approaches because they make more complete use of the available information. Such models can be trained on more basic inputs and learn complex correlations. Nevertheless, in some cases, we *explicitly do not* want the model to learn specific correlations or depend on certain features. The two key reasons are:

- we know that a given feature systematically differs between training and testing data,
- we want to use that feature in some downstream statistical procedure, whose quality degrades if the classifier changes the shape of the feature.

A simple solution might be not to use the problematic feature as input. However — given that machine learning is excellent in identifying correlations — this usually is not sufficient. Instead, several systematic *decorrelation* techniques exist.

A very versatile and practically straightforward approach is referred to as *planing* [136, 137]. Planing adds weights to the training data so that the resulting distribution in the relevant feature is either uniform or at least indistinguishable between different classes. Planing weights can be calculated by histogramming the feature separately for each class and using the inverse of the bin content as weight. The per-sample weights are then introduced by multiplication in the objective function. While planing is not guaranteed to achieve optimal results, its practical performance is typically good, making it an excellent baseline approach. Note that due to the explicit histogramming, planing becomes difficult in more than one or two dimensions.

Alternatively, one can modify the objective function when training a neural network to enforce independence between the network output and a specific feature. Merely adding an information-theoretic measure of dependence (such as the Kullback–Leibler divergence, discussed in Sec. 14.2) is possible but difficult to train as it either requires knowledge of the underlying distribution or explicit binning. Instead, the most generic and robust approaches either introduce a second — adversarial — neural network or a differentiable dependence measure.

The adversarial network [138, 139] receives the output of a classifier as input and has the task of inferring the feature of interest. The idea is that if the feature can be extracted from the classifier output, then the output is certainly not independent from the observable. By adding the objective term of the adversary with the opposite sign, compared to the classifier, a penalty is given to classifier outputs that show strong dependence. Alternatively, it is also possible to decorrelate an intermediate representation of data, i.e. at a hidden layer of a classification network, instead of its final output.

Adversarial approaches achieve high-decorrelation but are delicate to train as a saddle-point instead of a minimum needs to be identified.[3]

Adding a term based on distance correlation is a recent alternative to adversarial training that yields similar performance [140]. It uses a measure of nonlinear dependence that is differentiable and simple to calculate using batches called distance covariance [141–143]. It is zero if and only if the variables (in this case, the variable to decorrelate against and the classifier output) are statistically independent; otherwise, it is positive. Therefore it can be added to the objective function and used as a regularization term. Its advantage compared to adversarial approaches is a more simple network training.

---

[3]We will return to the importance of jointly training networks for complex objectives in Chapter 18.

13.3 Adversarial attacks

Finally, we can investigate so-called *adversarial attacks*. These should not be confused with *adversarial training* in the previous section. Here one should think of the adversary as an entity with access to an already-trained network. The adversary aims to find perturbations to the inputs that would maximally confuse the network. For example, the flipping of labels in a classification task is provoked [144]. These attacks are especially relevant for networks exposed to the outside world, such as user authentication for an online banking system, and less relevant for controlled scientific environments. In a sense, we can assume that nature is not *actively* training to fool us.

However, studying possible adversarial attacks can still be a useful diagnostic tool, expose systematic uncertainties, and provide an upper bound on the impact of certain perturbations [145]. While several ways to construct adversarial attacks exist, the key idea is to use the differentiability of the network to find for each input feature a modification that nudges the output in the desired direction.

13.4 Exercises

Exercise 13.1. In this exercise, we use a simple reweighting technique — often called planing — to decorrelate variables.

(1) Use the Notebook (Exercise 13.1 from the exercise page) to access the data and additional material

(2) We work with a simple dataset for a classification task. Draw all pairwise scatterplots for different features for signal and background separately. Also calculate the correlation coefficients. We want to train a classifier that is not correlated with feature 0 for the background.

(3) Implement a fully-connected network using all input features. What is the best accuracy it obtains? How much does placing increasingly stringent cuts on the classifier output shape feature 0 for the background?

(4) By how much do the accuracy and shaping change if you do not use feature 0 as input?

(5) Calculate weights that flatten the input distribution of feature 0, but still use it as input to the network. What are the training results?

Summary 13.1.

(1) In realistic applications, the assumption that training and testing data are drawn from the same distribution is often violated. Systematic differences can exist between the domain used to train a network and the application domain.

(2) We can either aim to model the response of networks under the uncertainties. In this case, we refer to an uncertainty that shrinks with the available training data as epistemic, and to one that does not as aleatoric. Extensions of architectures to account for both types of uncertainties exist.

(3) Ensembling refers to training multiple identical models on the same data and combining their predictions. Bayesian networks instead replace weights with probability distributions from which weight values are sampled during training and evaluation.

(4) Alternatively, models can be constructed to be robust or decorrelated against certain perturbations.

(5) Adversarial attacks are a useful method to find unexpected large sensitivity of a model output to small perturbations in the input.

Chapter 14

Revisiting objective functions

Scope 14.1. We consolidate our understanding of objective functions for training networks and show how the previously introduced mean squared error and cross-entropy are closely related to objectives known from fitting functions by maximum likelihood estimation. We investigate how networks learn to reproduce probability distributions of experimental data. Furthermore, we relate conditional probabilities to regression and classification tasks. We finally discuss various objective functions to emphasize their central role in building desired algorithms.

Objective quantities are well known in physics for fitting functions f to data. Such a *fit function represents a model chosen by physicists* with parameters to be adjusted to match the data. Working with neural networks is, in principle, not much different, except that the *network represents a flexible model by itself*, which is adjusted to data.

For creating network models, the objective (loss, cost) function is the control center. In this function, physicists encode their scientific questions to data. Together with the network architecture, the training data and procedure, this yields an algorithm tailored to answer the scientific questions at hand.

The objective function $\mathcal{L}$ (the $\mathcal{L}$ stands for loss) yields a scalar value to be minimized for optimizing the network parameters (Sec. 4.4). It is fascinating that fitting millions of parameters can be quantitatively evaluated by monitoring this single scalar only.

To take advantage of the gradient-based learning, the objective function has to be differentiable. This is mostly the case for problems encountered in physics. In other cases, differentiable approximations might be found [146]. Finally, we discuss problems that need to be solved without a differentiable objective function at the end of this chapter.

14.1   Fitting function parameters to data

We first recall two commonly used function-fitting methods with similarities to neural network optimization: The χ^2-fit in Example 14.1 and the likelihood-fit in Example 14.2. Then we elaborate further on the importance of objective functions for neural networks.

Example 14.1. χ^2-fit in one dimension.   Take a physics motivated function $f(x, \theta)$ with the parameter θ that we want to adapt to measured data. With typically Gaussian distributed uncertainties $\sigma_{y,i}$ of the k measured data points $y_i(x_i)$, we sum the squared residuals between model f and data to a single scalar quantity χ^2 that is the objective quantity:

$$\chi^2 = \sum_{i=1}^{k} \left(\frac{y_i(x_i) - f(x_i, \theta)}{\sigma_{y,i}(x_i)} \right)^2 \tag{14.1}$$

The parameter θ of the function is optimized to the data such that χ^2 becomes minimal. Furthermore, the χ^2 probability allows us to quantify how well the model can explain the data.

14.2   Reproducing probability distributions

Measurement data in physics are considered samples drawn from a probability distribution p_{data} underlying an experiment or a theoretical calculation. Our first goal is to find an objective function, such that a network is guided to reproduce the probability distribution of such a given data set [1]. We will encounter applications of this objective in Chapter 18.

Example 14.2. Likelihood-fit in several dimensions. Another well-known method for fitting a function $f(\mathbf{x}, \theta)$ with the parameter $\theta \in \mathbb{R}^m$ to measured data $\mathbf{x} \in \mathbb{R}^n$ is the maximum likelihood method, in which the product of the function values $f(\mathbf{x}_i, \theta)$ is evaluated for a set of k data points $\mathbf{x}_i$. The product yields the likelihood function L, which is maximized for the optimal choice of the parameters θ' (a one-dimensional visualization is shown in Fig. 14.1). The maximization procedure is sometimes abbreviated by arg max (meaning searching for the *arg*ument that *max*imizes the functional value):

$$L(\theta) = \prod_{i=1}^{k} f(\mathbf{x}_i, \theta) \quad \text{with} \quad \left. \frac{\partial L}{\partial \theta} \right|_{\theta = \theta'} = 0 \tag{14.2}$$

$$\theta' = \arg\max_{\theta} L(\theta) = \arg\max_{\theta} \prod_{i=1}^{k} f(\mathbf{x}_i, \theta) \tag{14.3}$$

For numerical reasons, the logarithm is used, yielding summations instead of products. Then the negative sum of the logarithms is defined as the objective:

$$F(\theta) \equiv -\ln L(\theta) = -\sum_{i=1}^{k} \ln f(\mathbf{x}_i, \theta) \tag{14.4}$$

$F(\theta)$ is then minimized with respect to the parameters θ:

$$\theta' = \arg\min_{\theta} F(\theta) = \arg\min_{\theta} \left(-\sum_{i=1}^{k} \ln f(\mathbf{x}_i, \theta) \right) \tag{14.5}$$

Function fitting with the likelihood method yields a conditional probability $p(\mathbf{x}_i \mid \theta)$ that, given parameters θ of the function f, the data can be described. Bayes' theorem then provides the desired *a posteriori* probability $p(\theta \mid \mathbf{x})$ that the model $f(\mathbf{x}, \theta)$ follows from the data $\mathbf{x}$:

$$p(\theta \mid \mathbf{x}) = p(\mathbf{x} \mid \theta) \frac{p(\theta)}{p(\mathbf{x})} \tag{14.6}$$

This requires prior distributions $p(\theta)$ for the parameters θ, often chosen flat. Moreover, the measured data are already recorded, such that $p(\mathbf{x})$ remains unchanged when the parameters θ are varied.

Note that for Gaussian uncertainties in the data, the χ^2 expression (14.1) yields the same distributions in parameter phase space (except for constants) as the negative log-likelihood function (14.4).

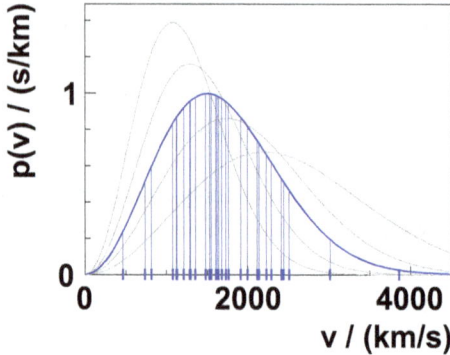

Fig. 14.1. Maxwell–Boltzmann probability density $p(v, T)$ for different gas temperatures T as a function of atomic velocity v (gray curves). A data set of measured velocities v_i is represented by the blue symbols on the horizontal axis. The blue curve maximizes the product of the vertical blue lines, i.e. gives the maximum likelihood (14.3) and hence the best estimate of the gas temperature T.

Network optimization using the likelihood method: The network $p_{\text{model}}(\mathbf{x}, \theta)$ represents our model with adjustable parameters θ. The desired network prediction $p_{\text{model}}(\mathbf{x})$ is to approximate the original data probability density $p_{\text{data}}(\mathbf{x})$ for any input value $\mathbf{x}$. Using k data points $\mathbf{x}_i$ from the data set, the negative log-likelihood function reads (14.4):

$$F(\theta) = -\sum_{i=1}^{k} \ln p_{\text{model}}(\mathbf{x}_i, \theta) \tag{14.7}$$

Follow the one-dimensional optimization sketch in Fig. 14.1. The measured atomic velocities v_i are compared with several models, in this example, Maxwell–Boltzmannn's probability distributions at various temperatures. The density of the measured velocities guides the likelihood procedure. If the model predicts large probabilities for velocity values v from intervals $v \in [v_1, v_1 + dv]$ containing many data, and low probabilities for regions $v \in [v_2, v_2 + dv]$ with a low data occupation, then the negative log-likelihood function becomes minimal.

Correspondingly, for the high-dimensional situation this optimization process reads:

$$\arg\min_{\theta} F(\theta) = \arg\min_{\theta} \left(-\sum_{i=1}^{k} \ln p_{\text{model}}(\mathbf{x}_i, \theta) \right) \tag{14.8}$$

Expectation values of probability distributions and Kullback–Leibler divergence: For comparing with the true probability distribution p_{data}, we normalize the negative log-likelihood function $F(\theta)$ by the number k of data values to obtain expectation values $\mathbb{E}$. To specify from which probability distribution the data points are drawn we write $\mathbf{x} \sim p_{\text{data}}$:

$$\mathbb{E}_{\mathbf{x} \sim p_{\text{data}}} \left[\ln p_{\text{model}}(\mathbf{x}, \theta) \right] = -\frac{1}{k} \sum_{i=1}^{k} \ln p_{\text{model}}(\mathbf{x}_i, \theta) \qquad (14.9)$$

Correspondingly, we evaluate the expectation value of the data distribution which we get directly from the data:

$$\mathbb{E}_{\mathbf{x} \sim p_{\text{data}}} \left[\ln p_{\text{data}}(\mathbf{x}) \right] = -\frac{1}{k} \sum_{i=1}^{k} \ln p_{\text{data}}(\mathbf{x}_i) \qquad (14.10)$$

A measure of the dissimilarity of two probability distributions is the Kullback–Leibler divergence $\mathcal{D}_{KL}$. It does not represent a metric distance measure,[1] but if the network model is identical to the probability density of the data, it gives $\mathcal{D}_{KL} = 0$:

$$\mathcal{D}_{KL}(p_{\text{data}} \| p_{\text{model}}) = \mathbb{E}_{\mathbf{x} \sim p_{\text{data}}} \left[\ln p_{\text{data}}(\mathbf{x}) - \ln p_{\text{model}}(\mathbf{x}, \theta) \right] \qquad (14.11)$$

$$= \mathbb{E}_{\mathbf{x} \sim p_{\text{data}}} \left[\ln \frac{p_{\text{data}}(\mathbf{x})}{p_{\text{model}}(\mathbf{x}, \theta)} \right] \qquad (14.12)$$

When we train the network model, the data is already given. Therefore, when we use expectation values for network training, we only need to minimize the expectation value of the model:

$$\arg \min_{\theta} \mathcal{D}_{KL}(p_{\text{data}} \| p_{\text{model}}) = \arg \min_{\theta} \left(-\mathbb{E}_{\mathbf{x} \sim p_{\text{data}}} \left[\ln p_{\text{model}}(\mathbf{x}, \theta) \right] \right) \quad (14.13)$$

Of course, this is the same result as we obtained using the likelihood-method (see (14.8)). Let us relate this to the cross-entropy objective function shown in (4.15). In general, an expectation value $\langle t \rangle$ given a probability density p is calculated by $\langle t \rangle = \mathbb{E}[t] = \int_{-\infty}^{\infty} p(x) \, t(x) \, dx$. For the expectation value (14.13) of $t = \ln p_{\text{model}}$ and discrete samples $\mathbf{x}$ from the probability distribution $p = p_{\text{data}}$, we recover the cross-entropy function (4.15). Thus, minimizing the Kullback–Leibler divergence $\mathcal{D}_{KL}$ (14.11) corresponds to minimizing the cross-entropy between two probability distributions and is equivalent to maximum likelihood estimation.

For learning probability distributions, we will introduce the Wasserstein or Earth mover's distance in Sec. 18.2.5. This provides a proper distance

---

[1] The Kullback–Leibler divergence is not symmetric in its arguments and does not satisfy the triangle inequality.

measure between two probability distributions. It can — for example via the so-called Kantorovich–Rubinstein duality — be approximated via an additional *critic* network and, in this case, conveniently computed using expectation values and enforcing a bound on the gradient of the objective (Lipschitz constraint).

Neyman–Pearson Lemma and Kullback–Leibler divergence: Many physicists are familiar with the likelihood ratio, which — according to the Neyman–Pearson lemma — has optimal properties for hypothesis tests of probability densities.

Let model p_1 have probability density $p_1(\mathbf{x})$ and an alternative model p_2 the density $p_2(\mathbf{x})$. An example is the velocity distribution of gas atoms at fixed temperature where model p_1 corresponds to the Maxwell–Boltzmann distribution and model p_2 to the Gaussian distribution. The likelihood ratio is defined by:

$$\lambda(\mathbf{x}) = \log \frac{p_1(\mathbf{x})}{p_2(\mathbf{x})} \tag{14.14}$$

With $\lambda > 0$, the data $\mathbf{x}$ are better described by model p_1 than by model p_2. For $\lambda < 0$, the description by model p_2 is better. For $\lambda \approx 0$ we cannot distinguish the two hypotheses. When designing such hypothesis tests, we want to achieve large separation power and thus large absolute values of λ. The Neyman–Pearson lemma states that for known densities, the likelihood-ratio is the optimal test statistic.

Let us calculate the expectation value for λ for data $\mathbf{x}$ distributed according to p_{data}:

$$\mathbb{E}_{\mathbf{x} \sim p_{\text{data}}}[\lambda(\mathbf{x})] = \mathbb{E}_{\mathbf{x} \sim p_{\text{data}}}[\log p_1(\mathbf{x}) - \log p_2(\mathbf{x})] \tag{14.15}$$

Thus, the likelihood ratio's expectation value corresponds to the Kullback–Leibler divergence (14.11) measuring agreement of probability distributions [147]. Put differently, a binary classifier trained using the usual cross-entropy objective function learns to approximate the likelihood-ratio between the two classes. We will encounter several applications of this *likelihood-trick* in later chapters.

Example 14.3. Reproducing Maxwell–Boltzmann's distribution: Here we consider the velocity distribution of gas atoms in one dimension. For a given fixed temperature $T_\circ$, the velocities v follow the Maxwell–Boltzmann probability density $p_{MB}(v, T_\circ)$ (Fig. 14.1).

Our network represents the desired model $p_{\text{model}}(v, \theta)$ with adjustable parameters θ and receives as input measured values of the atomic velocities v_i only. The parameters are optimized using the negative log-likelihood function $F(\theta)$:

$$\arg\min_\theta F(\theta) = \arg\min_\theta \left(-\sum_{i=1}^{k} \ln p_{\text{model}}(v_i, \theta) \right) \tag{14.16}$$

To compare with the true velocity distribution, we calculate the expectation value $\mathbb{E}$ of the network model by dividing by the number k of measurements:

$$\mathbb{E}_{v \sim p_{MB}} \left[\ln p_{\text{model}}(v, \theta) \right] = -\frac{1}{k} \sum_{i=1}^{k} \ln p(v_i, \theta) \tag{14.17}$$

We get the expectation value for the logarithm of the Maxwell–Boltzmann probability density from the data (or here analytically):

$$\mathbb{E}_{v \sim p_{MB}} \left[\ln p_{MB}(v, T_\circ) \right] \tag{14.18}$$

The dissimilarity of the two probability distributions is in terms of the Kullback–Leibler divergence:

$$\mathcal{D}_{KL}(p_{MB} \| p_{\text{model}}) = \mathbb{E}_{v \sim p_{MB}} \left[\ln \frac{p_{MB}(v, T_\circ)}{p_{\text{model}}(v, \theta)} \right] \tag{14.19}$$

For the minimization we only need the expectation value formed over the model

$$\arg\min_\theta \left(-\mathbb{E}_{v \sim p_{MB}} \left[\ln p_{\text{model}}(v, \theta) \right] \right) \tag{14.20}$$

together with the normalization requirement $\int_{-\infty}^{\infty} p_{\text{model}}(v, \theta)\, dv = 1$ (Fig. 14.1).

14.3 Conditional probability in supervised learning

Let us now rephrase two previously encountered problems in the language of maximum likelihood estimation: regression and classification.

Regression: Our goal is to estimate a conditional probability $p(y \,|\, \mathbf{x})$. We want to construct a network model $p_{\text{model}}(\mathbf{x}, \theta)$ with parameters θ that predicts a continuous value $\hat{y}_i = p_{\text{model}}(\mathbf{x}_i, \theta)$ given a data set with k data elements $\mathbf{x}_i$ [1].

Let one caveat be that uncertainties in the measurement data $\mathbf{x}_i$ lead to variations in the network predictions $\hat{y}_i$ which we collect in terms of a standard deviation σ_y. Thus, we specify the conditional probability $p(y \,|\, \mathbf{x})$ with a Gaussian with the target values y (labels) as the desired mean values, the model predictions $\hat{y}_i$ and a standard deviation σ_y:

$$p(y_i \,|\, \mathbf{x}_i) = \frac{1}{\sqrt{2\pi}\sigma_y} \exp\left(-\frac{(y_i - \hat{y}_i(\mathbf{x}_i, \theta))^2}{2\sigma_y^2}\right) \tag{14.21}$$

We again form the negative log-likelihood function:

$$F = -\sum_{i=1}^{k} \ln p(y_i \,|\, \mathbf{x}_i) \tag{14.22}$$

Because of the Gaussian, the sum simplifies to:

$$F = \sum_{i=1}^{k} \left(\ln\left(\sqrt{2\pi}\sigma_y\right) + \frac{1}{2}\frac{(y_i - \hat{y}_i(\mathbf{x}_i, \theta))^2}{\sigma_y^2}\right) \tag{14.23}$$

If the uncertainty σ_y is approximately constant we are left for the minimization challenge with the quadratic sum of the residuals between the target values y_i and the predictions $\hat{y}_i$ of the network:

$$\arg\min_{\theta} F = \arg\min_{\theta} \left(\sum_{i=1}^{k} (y_i - \hat{y}_i(\mathbf{x}_i, \theta))^2\right) \tag{14.24}$$

Apparently, the likelihood-based optimization of the conditional probability (14.21) corresponds to minimizing the mean squared error (MSE) objective function that we had introduced *ad hoc* in Sec. 4.4.1 (4.11):

$$\mathcal{L} = \text{MSE} = \sum_{i=1}^{k} (y_i - \hat{y}_i(\mathbf{x}_i, \theta))^2 \tag{14.25}$$

Note also that we train our network model directly for the posterior distribution $p(y \,|\, \mathbf{x})$. In the likelihood fit, this distribution was obtained only after applying Bayes' theorem (14.6).

Example 14.4. Prediction of gas temperatures: We measure the velocities v of m gas atoms at constant temperature T which we determine with a thermometer. The velocities of the gas atoms are measured by a high-precision method. We repeat the measurements at different temperatures to obtain k data sets $(T, \mathbf{v})$, each with a temperature measurement T and m velocity measurements captured by $\mathbf{v}$.

We, of course, know that the velocity data follow the Maxwell–Boltzmann distribution at the given temperature (Fig. 14.1). Nevertheless, we train a network $p_{\mathrm{model}}(\mathbf{v}, \theta)$ by inputting velocity data sets $\mathbf{v}_i$ and comparing the network predictions $\hat{T}_i = p_{\mathrm{model}}(\mathbf{v}_i, \theta)$ to the target values T_i of the thermometer. In doing so, we minimize the quadratic sum of the residuals:

$$\arg\min_{\theta} F = \arg\min_{\theta} \left(\sum_{i=1}^{k} \left(T_i - \hat{T}_i(\mathbf{v}_i, \theta) \right)^2 \right) \tag{14.26}$$

After successful network training, the network directly predicts the gas temperature $\hat{T}_i$ for each data set of velocity measurements $\mathbf{v}_i$.

Classification: We investigate the conditional probability $p(y \mid \mathbf{x})$ for the case of a binary classification problem. For this purpose, we can use the Bernoulli probability distribution $p_B(y \mid z)$ with $z \in [0, 1]$. It returns for the binary variable $y = 0, 1$ the probability that y is 0 or 1:

$$p_B(y \mid z) = z^y (1 - z)^{(1-y)} \tag{14.27}$$

Our network model $p_{\mathrm{model}}(x, \theta)$ is intended to reproduce the Bernoulli parameter z for a data set with input values $\mathbf{x}_i$ and target values y_i (labels), e.g. by designing a network that holds one output node for each class and ensuring the normalization of probabilities using the softmax function.

We compute the negative log-likelihood function and optimize the adaptive parameters θ using a data set of size k:

$$\begin{aligned} F &= \sum_{i=1}^{k} \ln \left(p_{\mathrm{model}}(\mathbf{x}_i, \theta)^{y_i} (1 - p_{\mathrm{model}}(\mathbf{x}_i, \theta))^{(1-y_i)} \right) \\ &= -\sum_{i=1}^{k} \left(y_i \ln p_{\mathrm{model}}(\mathbf{x}_i, \theta) + (1 - y_i) \ln (1 - p_{\mathrm{model}}(\mathbf{x}_i, \theta)) \right) \end{aligned}$$

$$\tag{14.28}$$

$$\hat{\theta} = \arg\min_{\theta} F$$

$$= \arg\min_{\theta} \sum_{i=1}^{k} \left(-y_i \ln p_{\text{model}}(\mathbf{x}_i, \theta) - (1 - y_i) \ln (1 - p_{\text{model}}(\mathbf{x}_i, \theta)) \right)$$

$$(14.29)$$

This is the cross-entropy objective function (4.16). Overall, we see here that cross-entropy is directly linked to maximum likelihood estimation.

14.4 Adaptive objective function

Here, we address new developments in the area of objective functions. We start with a remarkable approach of combining known candidate objective functions that can be adapted dynamically even during network training.

Again, we address the task of constructing the conditional probability $p(y \,|\, \mathbf{x})$ by a network model $p_{\text{model}}(\mathbf{x}, \theta)$ with parameters θ. In the previous section, we regarded the Gaussian distribution for $p(y \,|\, \mathbf{x})$ (14.21).

This yielded the mean squared error (MSE) as the objective function for optimizing the network parameters. In Fig. 14.2, the Gaussian distribution, the MSE, and its derivative are shown as a solid blue curve, where the variable denotes $x \equiv y - \hat{y}$, the difference between label and prediction.

Because of the *central limit theorem*, the Gaussian distribution is generally a good choice in physics, given all the uncertainties involved. But sometimes, many outliers exist, or other distributions are underlying a problem. An example is the Cauchy or Lorentz distribution for resonances, whose

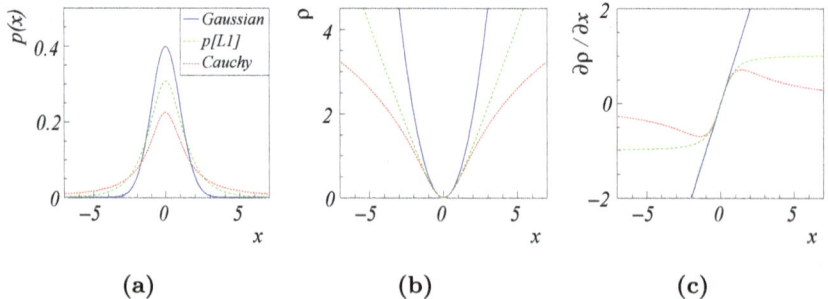

Fig. 14.2. Probability p, objective function ρ and its derivative $\partial\rho/\partial x$ for different values of the parameter α of the function ρ (14.30), Gaussian $\alpha = 2$, probability density of $L1$ $\alpha = 1$, Cauchy $\alpha = 0$.

probability distribution p together with the corresponding objective function and its derivative are also plotted in Fig. 14.2 as dotted brown curves. A third example is the probability density corresponding to the objective function $L1 = |x| = |y - \hat{y}|$ (dashed green curves).

Both objective functions reveal a flatter shape at large differences x, leading to smaller gradients than for the MSE. The avoidance of substantial gradients can be technically an advantage so that other objective functions can also be considered even for problems with underlying Gaussian distributions.

A quite general objective function was formulated in [148], which combines all important objective functions, so to speak, and can be adjusted dynamically during network training:

$$\rho(x, \alpha, c) = \begin{cases} \frac{1}{2} \left(\frac{x}{c}\right)^2 & \text{if } \alpha = 2 \\ \log\left(\frac{1}{2}\left(\frac{x}{c}\right)^2 + 1\right) & \text{if } \alpha = 0 \\ 1 - \exp\left(-\frac{1}{2}\left(\frac{x}{c}\right)^2\right) & \text{if } \alpha = -\infty \\ \frac{|\alpha - 2|}{\alpha}\left(\left(\frac{(x/c)^2}{|\alpha - 2|} + 1\right)^{\alpha/2} - 1\right) & \text{otherwise} \end{cases} \quad (14.30)$$

The parameter c is a scale parameter that can be used to adjust the extent of the region around $x = 0$. The parameter α continuously links the mentioned objective functions and probability densities (Fig. 14.2: $\alpha = 0$ Cauchy, Lorentz; $\alpha = 1$ probability density of $L1$; $\alpha = 2$ Gaussian). More objective functions and detailed justifications are shown in [148].

14.5 Hamiltonian gradient objective function

Conservation laws play a central role in physics. Therefore, it is natural to link the forecast of the time evolution of a physical state by a network model to a conservation law.

We consider a problem from classical mechanics — namely, the prediction of future trajectories from past observations. We know that in this system, the total energy is conserved. How can such crucial information be passed on and included when training a network? To solve this, we present an approach in which the Hamiltonian function H is learned by a network. The objective function is formulated via the *gradient* of the Hamiltonian function [149].

In Hamiltonian formalism we use the generalized coordinates q, p (position, momentum) to formulate an appropriate expression for the total energy of a physical system. The vector field $\vec{S}$ contains the partial derivatives of the Hamiltonian function H with respect to p and q and allows the calculation of the system state following in time:

$$\vec{S} = \begin{pmatrix} \frac{\partial H}{\partial p} \\ \frac{\partial H}{\partial q} \end{pmatrix} \tag{14.31}$$

$$\begin{pmatrix} q(t_1) \\ p(t_1) \end{pmatrix} = \begin{pmatrix} q(t_0) \\ p(t_0) \end{pmatrix} + \int_{t_0}^{t_1} \vec{S} \, dt \tag{14.32}$$

The network receives as input values the current coordinates q, p. It is supposed to learn the Hamilton function H from the phase space covered by the q, p values and predict the total energy E_{total}.

For the objective function we exploit that in the Hamilton formalism the partial derivatives $\partial H/\partial p$ and $\partial H/\partial q$ correspond to the time derivatives dp/dt and dq/dt:

$$\frac{\partial H}{\partial q} = -\frac{dp}{dt} \tag{14.33}$$

$$\frac{\partial H}{\partial p} = \frac{dq}{dt} \tag{14.34}$$

The time derivatives are calculated based on the training data. The derivatives of the Hamilton function (14.31) are obtained directly from the *back-propagation* method during training (Sec. 4.5). The correspondence (14.33) and (14.34) is enforced in the objective function by the mean-squared-error and in this way H is trained:

$$\mathcal{L} = \left(\frac{\partial H}{\partial p} - \frac{dp}{dt} \right)^2 + \left(\frac{\partial H}{\partial q} + \frac{dq}{dt} \right)^2 \tag{14.35}$$

Note that the network training proceeds indirectly. There are no labels for the value of H, but the network is supposed to infer from the gradients how the Hamiltonian function H should look like and which value of the total energy E_{total} is to be predicted. Such training is called *unsupervised*. A comparison of different training methods with and without labels is presented in Chapter 16.

When predicting the time evolution with the trained network, the network prediction H, partially differentiated in (p, q) (14.31), corresponds to the time derivatives $(dp/dt, dq/dt)$ according to (14.33) and (14.34). For obtaining the next time step, the time integral (14.32) is used. Further details on the Hamiltonian ansatz and examples are discussed in Example 14.5 and Ref. [149].

Example 14.5. The spring pendulum is solved according to Newton's equation of motion $m\, d^2x/dt^2 = -k\,x$. Using the frequency $\omega = \sqrt{k/m}$ and the generalized coordinates for position $q = x$ and momentum $p = m\, dq/dt$ we obtain:

$$q = \quad q_0 \, \cos\left(\omega\, t\right) \tag{14.36}$$

$$p = -m\,\omega\, q_0 \, \sin\left(\omega\, t\right) \tag{14.37}$$

Alternatively, the energy approach yields the corresponding Hamiltonian representing the total energy of the system and its partial derivatives:

$$H = \frac{1}{2}k\,q^2 + \frac{1}{2}\frac{p^2}{m} = E_{\text{total}} \tag{14.38}$$

$$\frac{\partial H}{\partial q} = k\,q = -\frac{dp}{dt} \tag{14.39}$$

$$\frac{\partial H}{\partial p} = \frac{p}{m} = \frac{dq}{dt} \tag{14.40}$$

For the input data to the network, pairs of values q, p from (14.36) and (14.37) are selected. For the objective function (14.35) and the next time step, value pairs of the corresponding time derivatives $dq/dt, dp/dt$ are also calculated. The network task is to learn the Hamiltonian function H. Therefore, for the objective function, its partial derivatives $\partial H/\partial p$ and $\partial H/\partial q$ are not calculated by (14.39) and (14.40) but via backpropagation.

Figure 14.3 shows the performance of the Hamiltonian network compared to a standard neural network approach. This network was trained to predict the pendulum state as usual with labels but does not appear stable. In contrast, the Hamiltonian approach preserves the desired phase space of generalized coordinates q, p (Fig. 14.3, left), thereby conserving the total energy E_{total} of the pendulum in at least 20 time steps (right).

14.6 Energy-based objective function

Occupation of energy states E_i in physics is typically distributed like $\sim \exp\left(-E_i\right)$ indicating that configurations with overall low energies are much more frequent than configurations with high energies. Transferred to neural network training, the exponential function $p(x) = \exp\left(-E(x)\right)$ guarantees positive probability values. Thus, maximizing probability for parameter optimization can be replaced by an energy-based objective function that is minimized.

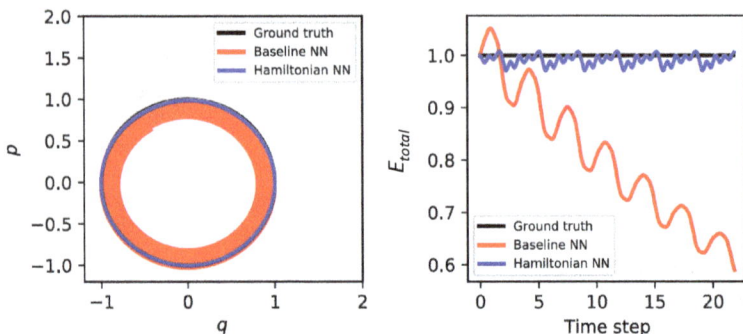

Fig. 14.3. Comparison of the time evolution of a pendulum predicted by the Hamiltonian neural network (blue) and a standard network (red) [149]: (left) generalized coordinates and (right) energy conservation as a function of time steps.

One application of energy-driven optimization is the so-called Restricted Boltzmann Machine (RBM) with the goal of *representation learning* (Fig. 1.1). For given data, a representation in the latent space of the network is to be found. Conversely, from the latent space, model data are to be generated that approximate the original data.

The minimal configuration is two network layers, the input layer with the data $\mathbf{x}$ and the hidden layer with the nodes $\mathbf{h}$ (Fig. 14.4). Inversely, the input layer $\mathbf{x}$ is also used as the output layer, so the network configuration corresponds to an undirected graph (Sec. 10.2). The keyword *restricted* denotes that connections exist exclusively between nodes $\mathbf{x}$ and $\mathbf{h}$ of the two layers, but not within a layer.

The connections of the layers are realized as usual with a weight matrix $\mathbf{W}$, a bias vector $\mathbf{b}$, and an activation function σ (compare Chapter 3). The output of the generated data uses the same weight matrix transposed as used for the forward step.

The setting of the parameters $(\theta = \mathbf{W}, \mathbf{b})$ is evaluated with a joint energy function, which is minimized as the objective function:

$$E_\theta(\mathbf{x}, \mathbf{h}) = -\sum_{i \in \text{visible}} a_i\, x_i - \sum_{j \in \text{hidden}} b_j\, h_j - \sum_{i,j} x_i\, h_j\, W_{ij} \qquad (14.41)$$

The probability of reproducing the data $\mathbf{x}$ by given values for $\mathbf{h}$ is:

$$p(\mathbf{x}) = \sum_h \frac{1}{Z_\theta} e^{-E_\theta(\mathbf{x}, \mathbf{h})} \qquad (14.42)$$

input/output hidden layer
layer

x h

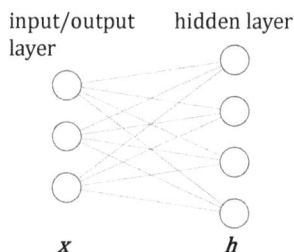

Fig. 14.4. Example of a Restricted Boltzmann Machine with the input/output layer and one hidden layer.

Here the partition function Z_θ is a normalization, representing all possible configurations.

The network training runs without data labels, i.e. it falls under *unsupervised learning*, which is explained in Chapter 16. While Restricted Boltzmann Machines are not necessarily deep networks themselves, they have influenced the development of Deep Learning over many years. For a detailed discussion refer to Ref. [150]. They are still further developed today, e.g. for many-body problems in quantum physics [151].

14.7   Learning without gradients

There are situations where gradient methods for training networks encounter limitations. An example is a network that is supposed to predict a discrete integer. The prediction is a step function that enters as an argument in the objective function for network training. However, the objective function cannot simply be partially differentiated by the network parameters. Similar issues occur when training a network to solve a problem where the solution involves many discrete steps, and the result is only known at the end. Examples from outside physics are finding a way out of a labyrinth or, rather famously, playing games [152].

Thus, an alternative approach is needed on how to optimize the network parameters. For this, a criterion for success needs to be formulated, and a (*policy*) how the parameters' phase space should be sampled. This form of learning has its own research field. It is referred to as *reinforcement learning*. Although several research papers have been published in physics, this area is beyond the scope of this textbook.

An application in proximity to reinforcement learning is mapping m discrete objects to n discrete target objects through a network. An example from particle physics is to relate m quarks and gluons emerging from a collision to the n particles observed in the detector (see Example 14.6).

Training the network requires a criterion for the correct mapping, which can be obtained from simulations. As a policy for sampling the possible phase space, different permutations in the mapping can be tried. While the network training is rather slow because of the permutations, the later mappings happen fast through the trained network.

Example 14.6. Policy: Particle collisions at the LHC can produce a Higgs particle together with a top quark and a top antiquark. These short-lived particles decay immediately, and the decay products are detected as clustered particle groups (jets) in the detector. Higgs decays into two jets and each top into three jets, so eight jets are measured. But which jet belongs to which parent particle? A network can be trained to predict this assignment [153].

The success criterion for the correct assignment between parent particles and jets in the detector can be formulated by simulated data of the particle collisions, in which the true relations are known. A permutation algorithm can be used as the *policy*. The final trained network then serves as a pre-stage to further analyze the measured data by a subsequent network, whose precision benefits from the assignment.

14.8 Different types of objectives

Finally, let us briefly consider the connection between "local" and "global" objectives of a machine learning algorithm. In this context, *local* means, for example, solving a given binary classification task between two processes using binary cross-entropy. However, this classifier output will usually not be the final scientific result but only an intermediate quantity and serves as input to downstream statistical analysis. This statistical analysis will also take other effects, such as measurement uncertainties, into account and yield the final, *global*, physical result.

If we set-up the machine learning task correctly, the local metric (e.g. accuracy for a classification task) will serve as a good proxy for the global metric (e.g. the precision of the measurement). Nevertheless, such an exact correspondence might not always be possible — especially in the presence of large uncertainties on the measurement. In such cases, it makes more sense to try to optimize the global goal immediately. This will involve finding differentiable approximations to a non-differentiable calculation and similar tricks to make the calculation tractable [146, 154].

Summary 14.1.

(1) Optimizing the parameters of a neural network using an objective function appears similar in many aspects to fitting a function using maximum likelihood estimation.

(2) The Kullback–Leibler divergence $\mathcal{D}_{KL}$ is a measure of the dissimilarity between the probability distribution underlying the data and the distribution predicted by the neural network model. The similarity of the distributions is indicated by $\mathcal{D}_{KL} = 0$, which can be directly related to minimizing the expectation value of the negative log-likelihood or minimizing the cross-entropy. Also, the expectation value for a likelihood ratio hypothesis test (Neyman–Pearson lemma) for the case of two indistinguishable distributions corresponds to $\mathcal{D}_{KL} = 0$.

(3) If in a regression task the data used for training can be assumed to have Gaussian uncertainties, then maximum likelihood-based parameter optimization is directly related to minimizing the mean squared error function.

(4) Training a classifier translates to learning a conditional probability distribution over the classes. For predicting a binary variable with $y = 0$ or $y = 1$ based on the Bernoulli distribution, the parameter optimization corresponds to minimizing the cross-entropy.

(5) Choosing a different loss function can be helpful, especially for outliers which induce substantial gradients at large differences between data labels and network predictions. Because of its correspondence to the Gaussian probability distribution, the mean squared error function is a good initial choice for physics problems. Nevertheless, using an objective function with dynamic adaptations of probability distributions, including Cauchy and Gaussian distributions, may be beneficial for some applications.

(6) To obtain the total energy in predicting an evolving physical state, the Hamiltonian function $H(p, q)$ can be learned by a neural network. The mean squared error on the gradients of the input data $(dp/dt, dq/dt)$ and the gradients $\partial H/\partial p, \partial H/\partial q$ obtained from backpropagation are used as the objective function. The network prediction on the time development of p, q results

from the correspondence of the partial derivatives following the Hamilton formalism.

(7) An energy-like optimization of the network parameters is realized in so-called Restricted Boltzmann Machines (RBMs). Their goal is learning representations of given data. The minimal configuration is an undirected graph consisting of two layers, exclusively with links between the layers. The first layer serves as input and output layer.

(8) For optimization problems without gradients of an objective function, new challenges arise, opening a new field of research referred to as reinforcement learning. One example is the prediction of a discrete output value. This field is beyond the scope of this textbook.

(9) It is important to consider what the final scientific question one wants to answer is, and if it can be — perhaps with suitable approximations — directly formulated as an objective function.

PART 4

Deep Learning Advanced Concepts

The following advanced neural network concepts convey the power of modern machine learning for advancing physics research. Instead of optimizing via data labels, the network is assigned functional goals for which it optimizes itself autonomously through the training process.

Chapter 15

Beyond supervised learning

In earlier parts of this book, we discovered how to solve a set of relatively well-defined tasks — classification and regression — with increasing sophistication and precision. For these problems, the objective function or loss function[1] was defined using independently known truth labels for the training data. We call such tasks *supervised* as there is additional information available. Of course, for data we finally apply the algorithm to, this information is missing.

In the following, we will explore what can be learned if such labels are either completely absent (*unsupervised learning*) or defined only on average or available for a subset of events (*weakly-supervised learning*).

However, there is still a lot that can be done with rare or absent labels. In Chapter 16, we first introduce training in the absence of per-example labels. The remainder of this part then introduces several advanced methods.

We encounter autoencoder networks in Chapter 17 and see how they can learn to compress data and reduce noise. In the absence of labels, we can also attempt to learn the underlying probability distribution of data examples. Once such a distribution is learned, we can sample from it to produce more data examples. In Chapter 18, three basic architectures for this task will be introduced: Generative Adversarial Networks (GANs), Variational Autoencoders (VAEs), and normalizing flows.

Whenever simulations are used — either in the form of generative machine learning models, or based on Monte Carlo sampling — the question arises how well the simulation describes the data. We will review so-called domain adaptation methods which aim to improve the agreement of distributions in Chapter 19.

---

[1] In this part, we follow the most frequent use in modern deep learning literature: loss function or simply loss.

Once we have models that learn the probability distribution of data, these can also be used to detect outliers, over-densities, and other types of anomalies. Several types of approaches, again including architectures like autoencoders and normalizing flows, will be introduced in Chapter 20.

Finally, in Chapter 21, we mention concepts that go beyond this textbook's scope and discuss some future perspectives in machine learning.

Chapter 16

Weakly-supervised classification

Scope 16.1.

Problems with perfect labels are all alike; every weakly labeled problem is imperfect in its own way. (Loosely after L. Tolstoi's *Anna Karenina*.)

This chapter introduces concepts of how learning can be possible with less-than-optimal labels. We will distinguish between three scenarios: incomplete, inexact, and inaccurate supervision and see how to accordingly adapt known tools from fully-supervised problems to these new challenges.

In the realm of supervised learning, a label y_i was provided for each data point $\mathbf{x}_i$. Weakening this requirement opens up several potential scenarios and new practical applications for classification. These, for example, can be grouped [155] into:

- Incomplete supervision,
- Inexact supervision, and
- Inaccurate supervision.

In the case of *incomplete* supervision, only a small subset of examples is labeled. Creating labels may be expensive and often requires human labor or dedicated experiments. In Sec. 16.1, we discuss two learning strategies for such problems.

With *inexact* supervision, we instead face the problems of overly coarse labels. For example, in drug discovery, we can measure that a given molecule produces the desired effect (binding to a target site). However, as molecules can adopt multiple shapes (e.g. by rotations), it is difficult to infer *which* shape works. In Sec. 16.2, we develop strategies to solve such problems.

Another case of weak supervision occurs when labels are defined for each example but might be *incorrect* — for example, through measurement errors. These cases are analyzed in more detail in Sec. 16.3. Mislabeling also introduces uncertainties on a result and, in this context, also relates to Chapter 13.

Finally, training to detect anomalies provides a further example of (very) weak supervision as labels for yet unknown anomalies can usually not be provided (Chapter 20).

16.1 Incomplete supervision

In classification problems with incomplete supervision, we have N data-points x_i of which only $L < N$ are assigned a true class label y_i. For the remaining $N - L$ points, no label is known. However, we can still assume that these points belong to one of the classes.

Practically, this can occur if the labeling process is slow, expensive, or otherwise difficult to carry out — for example, if the initial labeling needs to be done by human experts.[1] That labeling is often expensive is not surprising. If the task were easy or already automated, there would be no need to train a complex deep learning algorithm to solve it.

We can distinguish two possible scenarios: either the algorithm is allowed to query a so-called *oracle* (i.e. a human expert) for additional labels in the training process, or it is not.

If an oracle is available, we speak of *active learning* [156]. We assume that requesting additional labels is a costly operation, so the algorithm should learn to identify the data points for which obtaining a label would have the largest benefit. In scientific applications, this occurs if we can, in principle, carry out a measurement or complex calculation — such as obtaining a potential energy surface [157] — but it is computationally costly. We want to minimize the number of times it has to be done.

An example of an active learning algorithm is the *query by a committee* method: Several models are trained (in the usual way) on the available labeled L examples. The point for which these models have the most considerable disagreement in their output is then queried next.

Alternatively, if no oracle is available and only the initially labeled data points can be used, the problem is known as *semi-supervised learning*. Naively, one could use the available labeled example for training a fully-supervised algorithm and ignore the others.

---

[1]The Galaxy Zoo project http://www.galaxyzoo.org/ is a successful example of crowd-sourcing the classification of scientific data, in this case, images from telescopes.

However, under certain conditions — often met by realistic data — it is possible to profit even from using unlabeled data. These assumptions are [158]: smoothness (closeness of points implies closeness of labels), clustering (points belonging to the same cluster share a label), and the manifold assumption (the actual dimensionality of the data is much lower than the input space dimensionality — see also Example 18.1).

Given its large practical relevance, many semi-supervised learning approaches exist (see [158] for an extensive discussion). Generative models, for example, which we will encounter in Chapter 18, enable unsupervised learning of the probability density of data. This learned density can then be used for predictions.

Recently, progress has been made in using graph-based algorithms (see Sec. 10.1) for semi-supervised learning. Here the data points are represented by nodes in a graph, and their distance is encoded in the edges. Using the graph and labeled points, algorithms can exploit the structure of the data to infer additional labels. Note that this is a so-called *transductive* problem — the points to be labeled are already part of the training data.

16.2 Inexact supervision

We speak of inexact supervision when labels are not available with the granularity we desire. We might, for example, want to identify which *patch* of an image corresponds to a cat but are only given the information whether an image *as a whole* contains a cat or not as our training data.

The classical framing of this task is called *multiple-instance learning* [159]. Formally, we consider bags (corresponding to whole images) $\mathbf{x}_i$ of data with per-bag labels y_i as inputs to our algorithm. Each bag consists of a set of instances $\mathbf{x}_{ij}$ (the patches). The per-bag label is positive if any of the bag's instances is positive (contains the cat). While this setup might sound contrived, it often occurs in scientific and medical applications [159–162].

An overview of multiple-instance learning approaches is given in [163]. Three classes of algorithms exist: aggregating per-instance responses, focusing on bag level predictions, and embedding per-bag information into a feature vector so that classical supervised methods can be used.

A specific inexact supervision problem occurs in particle physics when simulation tools can statistically predict the composition of a given batch of experimental data. We then know the composition of a given batch, but not the individual label for each data point [164]. In such a case the loss function can be modified from its fully supervised form (see also (4.11)):

$$\mathcal{L}_{\mathrm{MSE}} = \frac{1}{b \cdot m} \sum_{j=1}^{b} \sum_{i=1}^{m} \left[f(\mathbf{x}_{ij}) - z(\mathbf{x}_{ij}) \right]^2 \qquad (16.1)$$

to a weakly-supervised version:

$$\mathcal{L}_{\mathrm{WMSE}} = \frac{1}{b} \sum_{j=1}^{b} \left[\frac{1}{m} \sum_{i=1}^{m} f(\mathbf{x}_{ij}) - \bar{z}_j \right]^2 \qquad (16.2)$$

The index j runs over b batches and i over m examples per batch in both cases, and $f(\mathbf{x}_{ij})$ is the network prediction for an individual example. In the fully-supervised case, we have per-example labels $z(\mathbf{x}_{ij})$. In the weakly supervised case, these are replaced with an average per-batch label $\bar{z}_j$ and compared with an averaged network prediction over the batch. While shown here for the mean squared error loss function, weak versions can also easily be obtained for other loss functions, such as the cross-entropy used in classification.

16.3 Inaccurate supervision

Finally, we turn to inaccurate supervision or the problem of label noise. In this case, a label is available for each data point, but it might not always be correct. In general, inaccurate labels will reduce the best achievable performance as well as the learning efficiency. Three broad strategies exist to mitigate this issue [165]:

- Developing classifiers that are inherently robust against label noise,
- Cleaning or filtering the training data to remove potentially incorrect labels,
- Learning to model the label noise explicitly.

Consider a specific example from nuclear physics [166]: distinguishing between neutron-induced and gamma-ray-induced pulse shapes in scintillation detectors. While the problem is well-defined, it is in practice difficult to record isolated training samples due to irreducible gamma-ray emissions alongside neutrons in fission processes.

Formally, we want to distinguish between two distributions given by their respective densities $p_0(\mathbf{x})$ (neutron-induced signals) and $p_1(\mathbf{x})$ (gamma-ray induced signals) but only have corrupted samples available for training:

$$\bar{p}_0(\mathbf{x}) = (1 - \alpha_0)\, p_0(\mathbf{x}) + \alpha_0\, p_1(\mathbf{x}) \tag{16.3}$$

$$\bar{p}_1(\mathbf{x}) = (1 - \alpha_1)\, p_1(\mathbf{x}) + \alpha_1\, p_0(\mathbf{x}) \tag{16.4}$$

Here $0 \le \alpha_0 < 1$ and $0 \le \alpha_1 < 1$ denote the amount of contamination, and the classes are chosen such that the majority of labels can be assumed[2] to be correct: $\alpha_0 + \alpha_1 < 1$.

We already know — see Sec. 14.2 — that a likelihood-ratio test is an optimal classifier. It can be shown that for each threshold μ, chosen so that

$$\frac{p_0(\mathbf{x})}{p_1(\mathbf{x})} > \mu \tag{16.5}$$

there exists another threshold

$$\lambda = \frac{\alpha_1 + \mu\,(1 - \alpha_1)}{1 - \alpha_0 + \mu\alpha_0} \tag{16.6}$$

so that

$$\frac{\bar{p}_0(\mathbf{x})}{\bar{p}_1(\mathbf{x})} > \lambda. \tag{16.7}$$

This is a beneficial result as it tells us that models trained on contaminated samples can learn equivalent decision functions to ones trained to pure samples, however, with a different calibration. Put differently, the specific threshold value to, e.g. obtain a 10% true-positive rate will come out wrong when using a contaminated sample, but the learned classifier is still useful. The above observation also forms the core of the classification without labels (CWoLa) approach [167] that we will encounter again when discussing anomaly detection in Chapter 20.

---

[2]Strictly speaking, the distributions also need to be sufficiently different or *mutually irreducible* which we can assume for practical problems. See Ref. [166] for a more detailed discussion of this assumption.

16.4   Exercises

Exercise 16.1. Zachary's karate club: In this example, we investigate semi-supervised node classification using Graph Convolutional Networks on Zachary's Karate Club dataset introduced in Example 10.2. Sometime ago there was a dispute between the manager and the coach of the karate club which led to a split of the club into four groups. Can we use Graph Convolutional Networks to predict the affiliation of each member given the social network of the community and the memberships of only four people? Develop your solution starting from Exercise 16.1 from the exercise page.

Summary 16.1.

Weak supervision denotes a large space of possible generalizations of supervised learning tasks:

- Incomplete supervision: some labels are missing. Active learning denotes strategies that can query additional labels during the training phase. Semi-supervised learning methods incorporate labeled and unlabeled data.
- Inexact supervision: labels are too coarse. In this case, per-batch labels can be used instead.
- Inaccurate supervision: labels might be incorrect. However, training a classifier to distinguish the polluted samples is often a valid strategy.

Chapter 17

Autoencoders: finding and compressing structures in data

Scope 17.1. This chapter introduces autoencoders as elementary networks for unsupervised learning. The focus is on bottleneck structures with encoder and decoder parts separated by a middle ('hidden') layer of restricted dimensions. Encoder and decoder can be trained so that the hidden layer contains salient features of the input data, providing a useful, compressed representation of the data. Different categories of autoencoders are distinguished, including sparse, contracting and de-noising types.

17.1 Autoencoder networks

Autoencoders constitute another elementary type of network architecture, which works quite differently from the networks encountered in previous chapters. Their origins lie in dimensionality reduction and data compression, although their applications extend nowadays far beyond those areas [1]. It is probably not an exaggeration to call autoencoders the 'workhorses' of unsupervised learning [168]. To understand the underlying idea, let us start with the basic scheme of an autoencoder, depicted in Fig. 17.1. It consists of an input and an output layer with the same dimensionality and an encoder and decoder separated by a middle layer of reduced dimensions. The task of the network is to reconstruct the input $\mathbf{x}$ at the output $\mathbf{x}'$ using the mappings $\mathbf{h} = g(\mathbf{x})$ (encoder) and $\mathbf{x}' = f(\mathbf{h})$ (decoder), which at first may seem not very useful. However, to prevent the network from just replicating the input by learning the identity function, the input is passed by the encoder through a layer of restricted dimensions

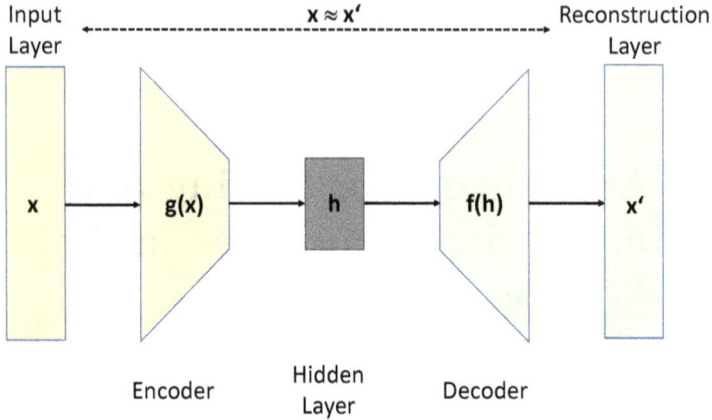

Fig. 17.1. Schematic basic autoencoder. The compressed representation **h** is also called latent variables, latent vector, latent space, or code.

before being reconstructed by the decoder. The autoencoder is then trained to minimize the reconstruction error using a suitable loss function

$$\mathcal{L}(\mathbf{x}, f(g(\mathbf{x})) = \mathcal{L}(\mathbf{x}, \mathbf{x}'), \tag{17.1}$$

which may be a simple mean squared error (4.12) for N-dimensional real-valued input:

$$\mathcal{L}_{MSE}(\mathbf{x}, \mathbf{x}') = ||\mathbf{x} - \mathbf{x}'||^2 = \frac{1}{N}\sum_{i=1}^{N}(x_i - x_i')^2, \tag{17.2}$$

or the cross-entropy (4.16) in the case of (properly normalized) discrete-valued input:

$$\mathcal{L}_{CE}(\mathbf{x}, \mathbf{x}') = \frac{1}{N}\sum_{i=1}^{N}[-x_i \log x_i' - (1 - x_i) \log (1 - x_i')] \tag{17.3}$$

Of course, a 'bottleneck architecture' provides an inherently lossy reconstruction of the input at the output. What makes autoencoders so valuable is that the middle layer contains a reduced representation of the data, which can even be used without the decoder once the training is completed [169]. This representation is commonly referred to as latent variables, latent vector, latent space, or code. The training is done by feeding the input to the output and forcing the reconstruction to resemble the input as much as possible via the loss function. This procedure can be performed in an unsupervised way since it requires no labeling of the input data. It is very attractive in the many cases where labeling is impractical or even impossible.

How are autoencoders able to provide highly compressed yet good representations of the input data? A key point is that typical real-world data are highly correlated internally. For example, photos contain millions of pixels with mostly gradual changes between neighboring pixels that hence can be compressed (also by traditional methods not involving machine learning). By contrast, since no magic is involved, an autoencoder cannot reconstruct anything meaningful from input with random changes from pixel to pixel. Consequently, de-noising of a blurred image is very well possible (see Sec. 17.2). In general, an autoencoder creates a representation of the data by mapping them onto a lower-dimensional surface. The use of nonlinear activation functions in the hidden layers and other parts of the network is another key issue enhancing the representational capability of the network. This is intuitively plausible, as one can imagine a warped hypersurface passing more easily through a manifold of points than its linear counterpart.

17.2 Categories of autoencoders

Autoencoders can be designed in many variants to ensure specific properties of the data representation. Important variants are *sparse, contractive, de-noising*, and *variational* autoencoders, which are discussed together with common architecture designs in the following. Being an alternative to generative adversarial networks, variational autoencoders are discussed in Chapter 18.

Shallow versus deep: In its simplest form, an autoencoder consists only of an input layer, one constricted middle layer, and an output layer. This is called a *shallow* autoencoder, the encoding is just performed by the dimensionality reduction from the input layer to the hidden layer without an extra layer being involved, and the same applies to decoding. However, deeper variants have turned out to be much more powerful since a larger number of layers also enhances the representational capability. A hierarchical data reduction is provided in a *deep* autoencoder as a by-product, as the representation in the most restricted layer is hierarchically related to those in intermediate layers.

Figure 17.2 shows an example of a deep autoencoder constructed symmetrically between the input and the output. Even if here only three layers

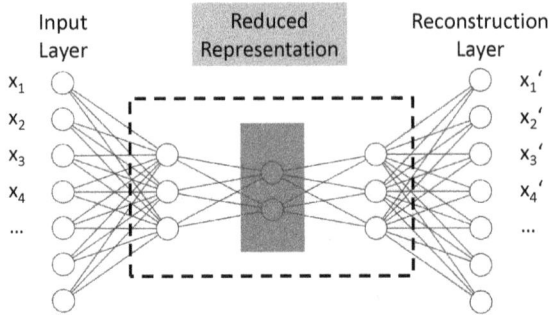

Fig. 17.2. Deep autoencoder with three hidden layers (dashed box).

are shown, autoencoders can be trained successfully with many more layers. Although symmetry is not a requirement, this type of architecture is frequently used, as it allows to share the weights between the kth layer and the $(M - k)$th layer for an M-layer architecture, which can reduce the training effort [168]. It should be emphasized that the apparent structural symmetry in Fig. 17.2 is to be taken with caution due to the nonlinearity of the activation functions associated with the layers. In general, encoder and decoder are two independent networks and not exact mirror-copies of each other. Specific choices of the activation functions may further be used to tailor the data compression to characteristic properties of the input.

Undercomplete versus overcomplete: So far, we have discussed a compressed representation of the data. This type of representation is called *undercomplete*, given the enforced dimensionality reduction by a constricted hidden layer. However, also *overcomplete* representations are possible, where the dimension of the hidden layer exceeds that of the input (see Fig. 17.3). This may appear of little use since, in theory, an overcomplete hidden layer could learn to copy the input with zero loss. However, this does not happen in practice, owing to a training process where weights are learned under constraints like regularization and/or sparsity (see Sec. 5.4 and below). For example, employing stochastic gradient descent in training is enough to mix the input at the hidden representation, thus providing an effective probabilistic regularization that prevents the network from learning trivial weights. In that sense, also overcomplete autoencoders are able to learn essential features of the data.

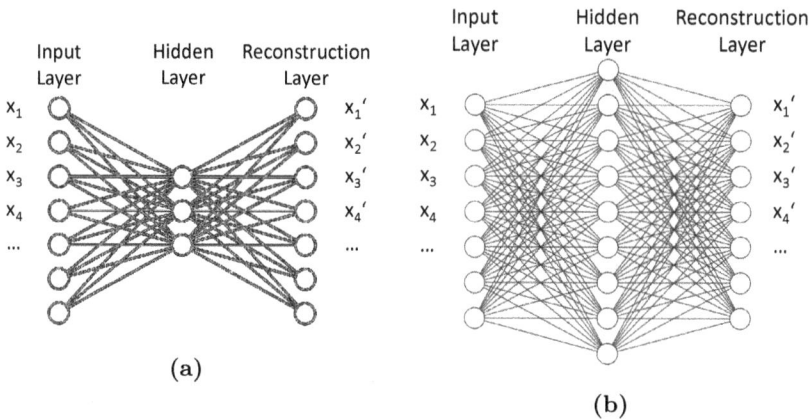

Fig. 17.3. Undercomplete (a) and overcomplete (b) autoencoders.

Regularized: This type of autoencoders is frequently used for classification tasks. It is constructed by imposing an additional constraint on the loss function.

In the case of *sparse* autoencoders, this is realized by a sparsity term acting on the compressed representation $\mathbf{h}$, which is added to the reconstruction error:

$$\mathcal{L}(\mathbf{x}, f(g(\mathbf{x}))) \rightarrow \mathcal{L}(\mathbf{x}, \mathbf{x}') + \lambda \Omega(\mathbf{h}) \tag{17.4}$$

Sparsity penalties can be formulated mathematically in various ways (see Chapter 14 of Ref. [1]). They result in the activation of only a small number of hidden units while deactivating the rest during training. This creates a highly effective bottleneck, even in overcomplete architectures.

The idea of the *contractive* autoencoder is to map a neighborhood of input points to a smaller region of output points (hence the name), i.e. the autoencoder is locally contractive with respect to perturbations of a training point [1]. This is achieved by employing derivatives as a specific penalty term:

$$\Omega(\mathbf{h}) = \sum_i ||\nabla_x h_i||^2 \tag{17.5}$$

The use of derivatives fosters the learning of features that change only slightly with varying input. The penalty term is only applied to training data, resulting in an encoding of salient features of the training distribution.

De-noising: Autoencoders of this type do not follow (17.1), where the minimization of the loss function ensures that the output $\mathbf{x}'$ is similar to the input $\mathbf{x}$. By contrast, the goal here is to obtain a representation insensitive to input distortions through a modified reconstruction criterion. This is achieved by minimizing instead

$$\mathcal{L}(\mathbf{x}, f(g(\tilde{\mathbf{x}})), \tag{17.6}$$

where $\tilde{\mathbf{x}}$ is a distorted ('corrupted') copy of $\mathbf{x}$. Whereas basic autoencoders aim to replicate the input, de-noising autoencoders strive to undo a corruption, taking advantage of the robustness of high-level representations with respect to corrupted input [170]. Such corruptions may be noise, but also other distortions like partly missing data.

All encoders remove noise when compressing data. However, de-noising autoencoders can be trained to remove *particular forms of noise* by using such noise in the training process. The training data are intentionally corrupted by adding noise characteristic for the respective field of application to the original input data. Since the loss is computed for the uncorrupted version of original data, the autoencoder learns to reconstruct largely correct representations from corrupted input.

How can we understand the de-noising process intuitively? In essence, the autoencoder uses the noisy input during training to assess the true manifold from which the input data originate. When finally being applied to data not encountered before, the autoencoder treats all input as corrupted, e.g. it attempts to project each point of the input data onto the true manifold as estimated from previous training. Consequently, the predicted origins of noisy input data will be those points on the manifold closest to the respective projections, as illustrated in Fig. 17.4. For a detailed treatment, the reader is referred to Chapter 14 of Ref. [1].

Since noise must be explicitly included in the training data (while already being present in the application data), the training is much facilitated when uncorrupted data are readily available, which is frequently the case.

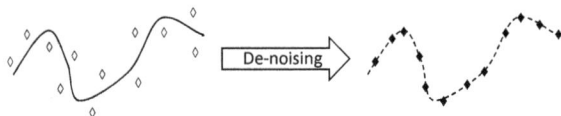

Fig. 17.4. De-noising autoencoder with corrupted input (open symbols) originating from a true manifold (solid line). The output (full symbols) results from a projection of the input onto an estimate of that manifold (dotted line). After Ref. [168].

For example, an autoencoder for de-noising photos could be trained using a set of high-quality images alongside their artificially blurred versions. Different forms of injected noise are common, depending on the type of input:

(a) For real values, Gaussian noise with zero mean and variance ζ is used;

(b) For discrete values, a fraction ρ of the input is randomly set to zero (masking noise).

Especially (b) reminds us of the dropout procedure from Sec. 5.4, and in fact, noise injection can be considered a form of regularization with ζ and ρ as regularization parameters. This works so well that overcomplete de-noising autoencoders are not unusual.

The architectures of autoencoders in real-world applications may be quite complex. For example, additional shortcuts have been introduced in autoencoder networks, which connect layers of the encoder and the decoder. An example architecture is UNet, which was introduced in Sec. 8.4.1. The purpose of shortcuts is to pass on information represented by the corresponding layers from the encoder to the decoder part. Such connections help to retain hierarchical feature representation, thus addressing the problem of vanishing/exploding gradients in deep networks. Shortcuts support back-propagation and can be essential for successful training (compare with Secs. 7.5 and 8.4.1).

17.3   Application areas

Applications of autoencoders are numerous. From a mathematical point of view, reduced representations of data are intimately related to the problem of finding relevant subspaces of multidimensional data. This task is known in linear algebra as Principal Component Analysis (PCA). It can be shown that a *linear* undercomplete autoencoder will generate the same subspace as PCA when trained with (17.2). Since autoencoders in deep learning utilize activation functions, they can be considered nonlinear generalizations of PCA [1].

Visualizations of multidimensional data in two or three dimensions are desirable in many scientific contexts. It should be noted that this task can be accomplished by a deep autoencoder in which the most restricted hidden layer has only the required dimensions. The compressed representation can then be mapped onto a plane or into a cube to visualize the data. Such plots might be difficult to interpret as they refer to an abstract coordinate system, but they can be beneficial to recognize clusters in complex data.

The reconstructions of the decoder are generally only needed during training. However, there are applications based on the decoder output. For example, the detection of unusual events can be related to the reconstruction loss as a similarity measure. This can be used to detect outliers in experiments or at production lines automatically, but also to uncover fraud in credit card data. This is thoroughly discussed in Chapter 20. Decoders may also be used to create artistic adaptations of images, including deceivingly realistic renderings known as 'deepfakes'.

Autoencoders can also be used to improve the learning process itself, for example, by initializing weights in a network before the actual training (pre-training).

17.4   Physics applications

We present two examples: autoencoders for anomaly detection (Example 17.1) and for de-noising radio measurements of cosmic-ray induced air showers (Example 17.2).

Example 17.1. Search for anomalies: In high-energy physics, autoencoders can be used to search for anomalies in jets of collision experiments. Their task is to monitor jet substructure from decays of heavy resonances, searching for outliers that may signal new physics. This can be performed efficiently in an unsupervised way by comparing experimental patterns to the compressed representation of an autoencoder. A corresponding bottleneck architecture was suggested in Ref. [171]. Its architecture, depicted in Fig. 17.5, involves several convolutional layers in the encoder and decoder to facilitate the search for patterns in images. Anomaly detection is discussed in more detail in Chapter 20.

1@40x40   10@40x40   10@20x20 5@20x20   400 100   100 400   5@20x20 5@40x40 10@40x40 1@40x40

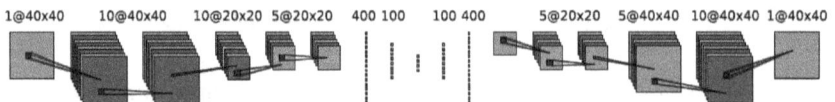

Fig. 17.5. Architecture of an image-based autoencoder for jet analysis in high energy physics. From [171].

Example 17.2. De-noising: The Pierre Auger Observatory for the detection of air showers from cosmic rays was already introduced in Example 11.2. The electromagnetic radiation generated in the atmosphere in the radio wave range is strongly contaminated by galactic noise as well as signals from human origin. Since these signals appear not to be very different from the background, the primary challenge in signal recovery is the disentanglement of signal and simultaneously recorded noise.

This problem was successfully addressed using a de-noising autoencoder network, which proved that de-noising autoencoders can be successfully employed for tasks with a very unfavorable signal-to-noise ratio [172]. Convolutional layers are used for searching translational invariant patterns in the time series, and the deep network allows to represent the input in terms of hierarchically related patterns. These representations are unfolded in the decoder part of the network and combined to recover the original dimension of the trace. During the training process, fine details in the characteristics of signal and background have been learned from the signal traces, and subsequently, only features associated with the signal are reconstructed. Results from the trained network are shown in Fig. 17.6. The network correctly identifies the weak signal in the noise-contaminated measurement and recovers its time evolution without appreciable noise.

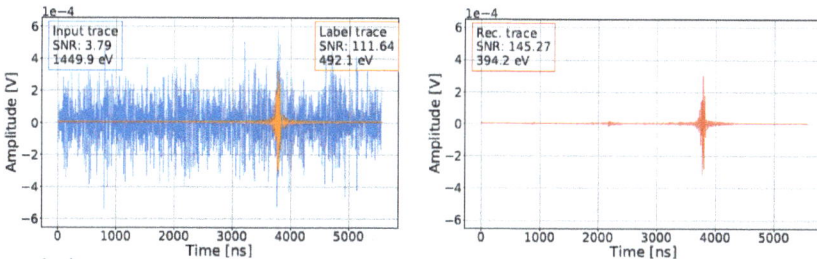

Fig. 17.6. Example of signal recovery and reconstruction by a de-noising autoencoder. (Left) The simulated signal (orange) is superimposed with noise (blue). (Right) Reconstructed signal (red). From [172] (©*IOP Publishing Ltd and Sissa Medialab. Reproduced by permission of IOP Publishing. All rights reserved*).

17.5 Exercises

Exercise 17.1. Speckle removal: Small-angle scattering of X-rays or neutrons enables insights into properties of nanostructured materials. In the case of X-rays from undulators at synchrotron radiation sources, extreme beam focusing can result in a high degree of coherence. Interference effects called *speckles* then appear naturally in the recorded data. This may be an unwanted effect that makes the measurements appear noisy. Starting from Exercise 17.1 from the exercise page, try to remove the speckles from the given test samples:

(1) Set up and train a deep convolutional autoencoder.
(2) State the test loss and comment on the reconstruction results for some test images.
(3) Apply the autoencoder to experimental data from partially coherent illumination and describe your observations.

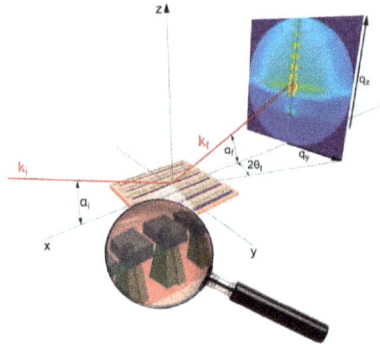

(a) Scattering geometry for a reflection experiment to access buried nanostructures. Reproduced from [173] with the permission of ©AIP Publishing.

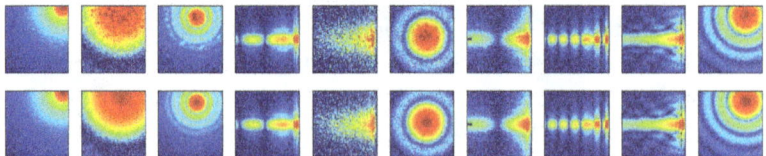

(b) Scattering patterns with and without superimposed speckles.

Fig. 17.7. Small-angle scattering of X-rays.

Summary 17.1.

(1) An autoencoder consists of an input and an output layer with the same dimensionality and one or more hidden layers. The encoder part of the network maps the input to a hidden representation, which is then mapped by the decoder to the output. This compressed representation is commonly referred to as latent space or latent variables. The autoencoder is trained to maximize the similarity between the input and the output. This allows unsupervised learning since the input is also fed to the output.

(2) Typically, the hidden layer is of reduced dimensions compared to the input (undercomplete autoencoder). Such 'bottleneck' architectures create reduced representations of the data, as they are inherently lossy. However, the dimensionality reduction enforces the learning of salient features of the input data.

(3) The decoder part is only needed for training and may be omitted once the training is completed. Nevertheless, there are applications based on the decoder output like outlier detection or generative networks.

(4) Deep autoencoders use multiple layers to create hierarchically reduced representations of the data. The representational power is further increased by the use of nonlinear activation functions in deep learning.

(5) Autoencoders exist in many variants; important categories are sparse, contractive, variational, and de-noising autoencoders. De-noising autoencoders learn to reconstruct clean versions from corrupted (noisy) inputs. Their training is based on the intentional inclusion of noise in the training data.

Chapter 18

Generative models: data from noise

Scope 18.1. Generative machine learning models — algorithms trained on some data that learn to produce additional realistic-looking examples — are versatile tools, both for scientific and non-scientific applications. We see how the three main architectures for building generative models — variational autoencoders, generative adversarial networks, and normalizing flows — address this challenging learning task and discuss how to train them efficiently. We conclude with some practical examples of applications in physics.

In the early parts of this textbook, we have largely dealt with discriminative models and supervised learning, i.e. the classification and reconstruction of objects. Typically, a sample $\mathbf{x}$ from a data set with probability distribution p_r was fed into a model to assign a label y. The models were trained to make accurate predictions, and their performance was evaluated using quality metrics like accuracy.

In the following, the creation of new samples is discussed by introducing the field of generative models. Instead of assigning labels, the task is to generate new samples that follow the data distribution p_r and thus are similar to the samples $\mathbf{x}$ used to train the model. For example, the goal would be to train a network using detector signals to generate new events that are not part of the training distribution. Hence, we aim to train a model to approximate the true distribution p_r with p_θ, where θ are the model parameters to be learned. After the training, we can use the approximation of our model to generate new samples $\tilde{\mathbf{x}} \sim p_\theta$ cost-efficiently. This basic concept is illustrated in Fig. 18.1. For the training of generative models, distance measures are utilized, i.e. quantities that capture the similarity between the true p_r and the generated distribution p_θ (compare with Sec. 14.2).

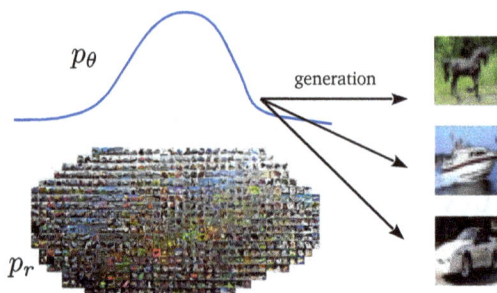

Fig. 18.1. Principle of generative models. The training distribution p_r is approximated using p_θ, which enables a fast and cost-efficient generation of new samples. Data from Ref. [174].

Several approaches of deep generative models exist, trying the challenging generation of new samples (see Example 18.1). One concept uses a low-dimensional representation of the data to enable a parameterization of the data distribution. By sampling from the parameterized distribution, new samples are created. This concept is called variational autoencoder (VAE) and is explained in Sec. 18.1.

Example 18.1. Image generation challenge: The generation of natural images is extremely challenging since the space of these images is high-dimensional. Imagine you want to generate new images of size $\dim(x) = 200 \times 200 \times 3$. To obtain meaningful samples using a random generator would be a lost cause, as we would generate noise only. Although, in theory, all real-world images are part of this (120 thousand) dimensional space, but obviously, these images take up only a small subspace.

The hypothesis that all the real-world images lie on low-dimensional manifolds y with dimension $\dim(y) \ll \dim(x)$ embedded within the high-dimensional space x is called the *manifold hypothesis*. For a detailed survey, refer to Ref. [175]. The success of deep generative models could then heuristically be explained by an adequate approximation of the low-dimensional manifold describing natural images.

A highly successful approach is generative adversarial networks (GAN). In this framework, instead of using a hand-coded loss function, a second *adversary* network is used. To generate new samples, a generator network is trained using the feedback of the second network, which evaluates the quality of the samples generated by the generator. Adversarial frameworks are discussed in detail in Sec. 18.2. Further, applications that deal with adversarial frameworks but do not belong to generative models are investigated in Sec. 19.1.

A third method is called *normalizing flows*, which employs *invertible networks*. In one direction, these networks learn the probability distribution underlying certain data. In the reverse direction, the networks generate new samples similar to the original data.

18.1 Variational autoencoder

The concept of Variational Autoencoders [176] (VAEs) is closely related to autoencoders discussed in Chapter 17. Here, we want to remind the reader that autoencoders are powerful methods to learn a compact and low-dimensional encoding of complex and high-dimensional data. As it is intractable to parameterize the data distribution in the input space directly, the goal of VAEs is to parameterize the data distribution formed in the low-dimensional encoding learned by an autoencoder.

Therefore, the basic architecture of VAEs is identical to an autoencoder and consists of two parts: an encoder and a decoder [see Fig. 18.2(a)]. While the encoder is trained to map the input data $\mathbf{x}$ into a compact representation often called *latent space*, the decoder is responsible for transforming the sample from this compact representation back into the input data space and output a reconstructed sample $\tilde{\mathbf{x}}$.

In natural images, the success of an autoencoder training is evaluated using the MSE (17.2) by comparing the reconstructed samples to the input samples. Here, the calculation is performed pixel-wise

$$\mathcal{L}_{\text{reco}}(x, \tilde{x}) = \frac{1}{N} \sum_{\mathbf{x}} ||\mathbf{x} - \tilde{\mathbf{x}}||^2, \tag{18.1}$$

where N denotes the batch size. For binary data, the cross-entropy can be used as reconstruction loss[1] (17.3).

To generate new samples after the training, it is essential to allow sampling in the latent space. Hence, VAEs explicitly constrain the distribution of the learned representation to follow a certain distribution p. By sampling from this distribution p, typically a Gaussian, and decoding the obtained

---

[1] In principle, other loss functions, e.g. adversarial losses [177] as used in GANs, are possible.

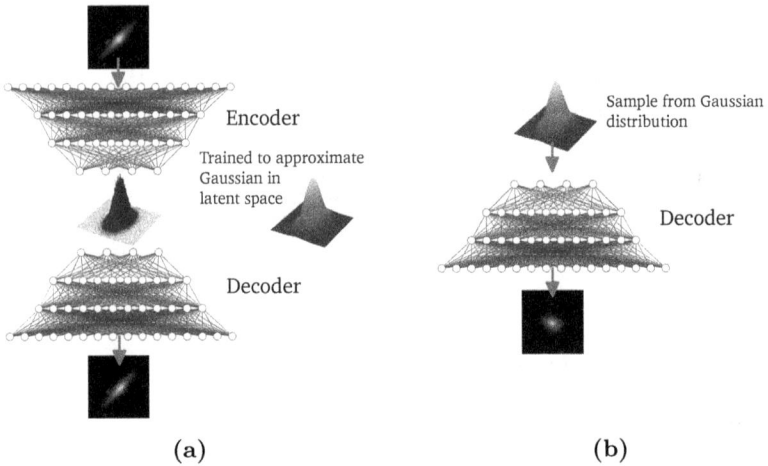

(a)   (b)

Fig. 18.2. Sketch illustrating the principle of variational autoencoders. Data from [179]. (a) Training of the variational autoencoder. The encoder is trained to approximate a Gaussian in the latent space, where the decoder should still be able to reconstruct the original image. (b) Sampling from a Gaussian and generation of new samples using the trained decoder.

samples using the decoder, new samples can be created. To constrain the representation, the loss function is extended and reads for a batch of size N:

$$\mathcal{L}(\mathbf{x}, \tilde{\mathbf{x}}, \mathbf{z}) = \mathcal{L}_{\text{reco}}(\mathbf{x}, \tilde{\mathbf{x}}) + \mathcal{L}_{\text{sim}}$$

$$= \mathcal{L}_{\text{reco}}(\mathbf{x}, \tilde{\mathbf{x}}) + \lambda \cdot \mathcal{D}_{KL}[q(\mathbf{z}|\mathbf{x})||p(\mathbf{z})] \tag{18.2}$$

$$= \mathcal{L}_{\text{reco}}(\mathbf{x}, \tilde{\mathbf{x}}) + \frac{\lambda}{N} \sum_{\mathbf{x}} \sum_{\mathbf{z}} q(\mathbf{z}|\mathbf{x}) \log \frac{p(\mathbf{z})}{q(\mathbf{z}|\mathbf{x})} \tag{18.3}$$

Here, λ is a hyperparameter to tune [178], and $\mathcal{D}_{KL}$ denotes the Kullback–Leibler divergence between the two distributions p and q (see Sec. 14.2 and Example 18.3). The desired distribution that the encoder should match is denoted by $p(\mathbf{z})$. The distribution $q(\mathbf{z}|\mathbf{x})$ is learned by the encoder to produce a representation $\mathbf{z}$ from an input $\mathbf{x}$. In principle, this added term in (18.3) can be interpreted as a measure of the similarity between the distributions p and q.

Usually, $p(\mathbf{z})$ is chosen to be a multivariate isotropic Gaussian. Therefore, instead of learning an arbitrary structure, the latent space consists of two vectors that describe a multivariate Gaussian. One vector $\boldsymbol{\mu}_z$ describes the mean values, and one vector the standard deviations $\boldsymbol{\sigma}_z$. With the so-called *re-parameterization trick*

$$\mathbf{z} = \mu_{\mathbf{z}}(\mathbf{x}) + \sigma_{\mathbf{z}}(\mathbf{x}) \cdot \epsilon, \ \epsilon \in \mathcal{N}(\mathbf{0}, \mathbb{1}) \tag{18.4}$$

a *vector encoding* $\mathbf{z}$ of the input sample is obtained by using the reconstructed μ_z, σ_z and noise ϵ sampled from a multivariate Gaussian. This procedure allows using backpropagation even if the model includes a stochastic sampling process. The obtained vector encoding $\mathbf{z}$ is used to reconstruct the input sample $\mathbf{x}$. Hence, the vector serves as input for the decoder and leads to a reconstructed sample $\tilde{\mathbf{x}}$.

Due to the assumption of $p(\mathbf{z})$ being an isotropic multivariate Gaussian[2] the final VAE loss function for a batch of N natural images reads:

$$\mathcal{L}(\mathbf{x}, \tilde{\mathbf{x}}, \mathbf{z}) = \mathcal{L}_{\text{reco}}(\mathbf{x}, \tilde{\mathbf{x}}) + \lambda \cdot \mathcal{D}_{KL}[\mathcal{N}(\mu_z(\mathbf{x}), \sigma_z(\mathbf{x}))||\mathcal{N}(\mathbf{0}, \mathbb{1})] \tag{18.5}$$

$$= \frac{1}{N} \sum_{\mathbf{x}} \left[||\mathbf{x} - \tilde{\mathbf{x}}||^2 + \frac{1}{2}(1 + \log \sigma_{\mathbf{z}}^2(\mathbf{x}) - \mu_{\mathbf{z}}^2(\mathbf{x}) - \sigma_{\mathbf{z}}^2(\mathbf{x})) \right]$$

and can be calculated analytically. Whereas the first part of the loss rates how well an image can be reconstructed, the second part measures how well the learned representation matches a multivariate Gaussian. λ has to be chosen by the user and is usually set to $\lambda = 1$ [176] or $\lambda > 1$ [178]. Note that for $\lambda \to 0$, the reconstruction quality generally increases. Still, since the representation no longer follows a Gaussian, the generated samples are of increasingly poor quality.[3] After a successful training, the encoder is no longer required [Fig. 18.2(b)]. Now the distribution formed by $q(\mathbf{z}|\mathbf{x})$ follows approximately a Gaussian. Hence, by sampling vectors from the multivariate Gaussian $\mathbf{z} \sim p(\mathbf{z})$ new samples $\tilde{\mathbf{x}} \sim p(\tilde{\mathbf{x}}|\mathbf{z})$ can be generated using the decoder.

18.2 Generative adversarial networks

In Generative Adversarial Networks (GAN) [180], a generator network g_θ is trained to approximate the true distribution p_r, where θ are parameters of the model to be learned. To create varying imitations of true images, the generator network g_θ is fed with noise variables $\mathbf{z} \sim p_z$ from the latent space. The output of the network is a generated image $\tilde{\mathbf{x}} = g_\theta(\mathbf{z})$. Thus, the generator network should learn the mapping from the latent space to the space of data samples (e.g. natural images, see Fig. 18.4) and thereby approximate p_r.

---

[2]For simplicity, the components of Gaussian are assumed to be independent.

[3]Sampling from the distribution formed by the training data in the latent space would create samples of good quality, which, however, would only be a replication of the input data.

Example 18.2. Generation of galaxy images: If an image of a galaxy is used to train a VAE, as depicted in Fig. 18.3, the encoder transforms the image into a Gaussian-like representation from which a vector is sampled. This vector represents the properties of the galaxy in the latent space. Using the vector encoding, the image of the galaxy is reconstructed using the decoder. By comparing the reconstructed and the input images, the reconstruction error can be determined. Finally, the 'similarity loss' is estimated by measuring the similarity between the estimated distribution and the multivariate Gaussian. After a successful training covering the iteration over thousands of galaxy images, the encoder is no longer required. By sampling from the multivariate Gaussian in the latent space, vector encodings can be obtained to generate new galaxy images using the decoder.

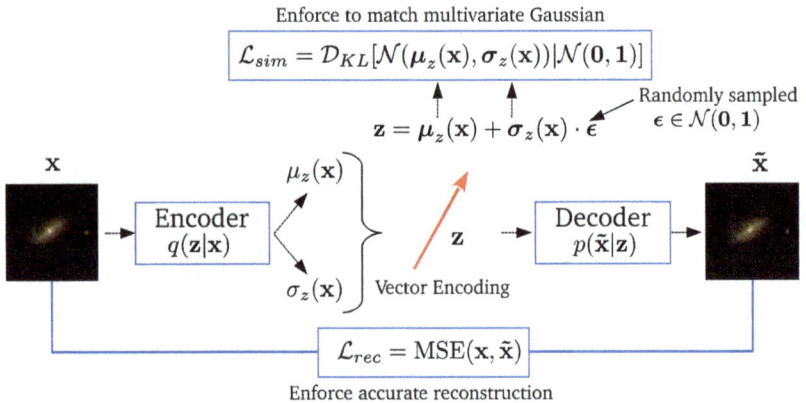

Enforce to match multivariate Gaussian

$$\mathcal{L}_{sim} = \mathcal{D}_{KL}[\mathcal{N}(\boldsymbol{\mu}_z(\mathbf{x}), \boldsymbol{\sigma}_z(\mathbf{x}))|\mathcal{N}(\mathbf{0}, \mathbf{1})]$$

Randomly sampled $\epsilon \in \mathcal{N}(\mathbf{0}, \mathbf{1})$

$$\mathbf{z} = \boldsymbol{\mu}_z(\mathbf{x}) + \boldsymbol{\sigma}_z(\mathbf{x}) \cdot \boldsymbol{\epsilon}$$

$\mathbf{x}$ $\mu_z(\mathbf{x})$ $\tilde{\mathbf{x}}$

Encoder $q(\mathbf{z}|\mathbf{x})$ $\mathbf{z}$ Decoder $p(\tilde{\mathbf{x}}|\mathbf{z})$

$\sigma_z(\mathbf{x})$ Vector Encoding

$$\mathcal{L}_{rec} = \text{MSE}(\mathbf{x}, \tilde{\mathbf{x}})$$

Enforce accurate reconstruction

Fig. 18.3. Training of a variational autoencoder using images of galaxies. Data from [179].

Latent space   Image space

Generator

$\mathbf{z} \sim p_z$ $\tilde{\mathbf{x}} \sim p_\theta$

Fig. 18.4. Principle of a generator network trained on the CIFAR-10 data set [174].

Example 18.3. The Kullback–Leibler divergence is a similarity measure between two probability distributions and is only 0 if $p(x) = q(x)$. The divergence states the extra amount of information needed to describe the same event when sampling from a distribution $p(x)$ instead of the underlying true distribution $q(x)$:

$$\mathcal{D}_{KL}(p\|q) = \mathbb{E}_{x\sim p}\left[\log p(x) - \log q(x)\right] = \sum_x p(x) \log \frac{p(x)}{q(x)} \qquad (18.6)$$

Since the expectation value is taken over p the KL divergence is not symmetric $\mathcal{D}_{KL}(p\|q) \neq \mathcal{D}_{KL}(q\|p)$. Especially note, wherever $p(x)$ is 0, the density of $q(x)$ does not matter (compare with Sec. 14.2).

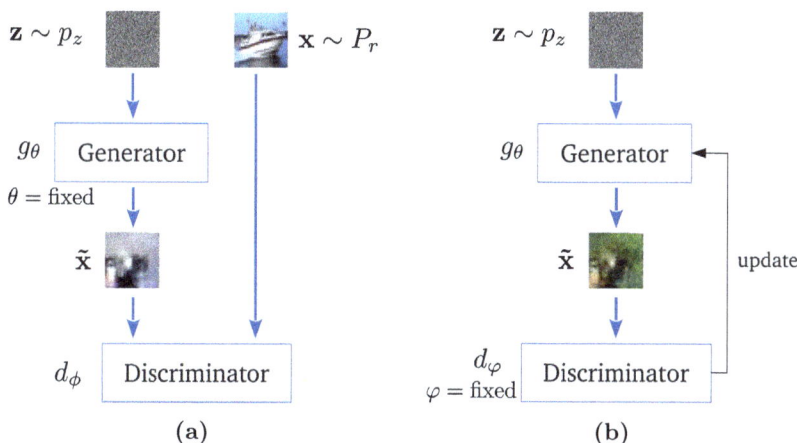

Fig. 18.5. GAN training on the CIFAR-10 [174] data set. (a) Discriminator training. The discriminator d is trained to classify between generated $\tilde{\mathbf{x}}$ and real $\mathbf{x}$ samples. (b) Training of the generator. The generator g is trained to fool the discriminator using its feedback.

It was always straightforward to train models in supervised learning, as we could evaluate their performance of our models using clearly defined objectives (cross-entropy, mean squared error). This is not the case anymore, as we want to approximate the data distribution to generate new samples. Hence, a measure of similarity between two distributions in a high-dimensional space is needed. Even if the evaluation of the image quality is easy for us humans, it is intractable to formulate this measure analytically. The ingenious idea is to perform this evaluation using a second *adversarial network*, the *discriminator network* d_φ, with trainable weights φ. This second network is a classifier trained to distinguish generated $\tilde{\mathbf{x}}$ and real samples $\mathbf{x}$. During the training of the generator, the discriminator should provide feedback if the generated samples are realistic or not. Combining both the generator and the discriminator network defines the adversarial framework.

18.2.1 *Adversarial training*

To train the generator using the feedback of the discriminator, *adversarial training* is used. This technique utilizes alternate updates of the discriminator and the generator. In the first step, the discriminator is trained to classify real and generated samples. In the second step, the generator tries to fool the discriminator, i.e. to generate samples that look realistic for the discriminator. In the classical GAN setup, each update covers only a single batch iteration.

Discriminator update: At first, the discriminator is trained to classify between real $\mathbf{x}$ and fake (generated) $\tilde{\mathbf{x}}$ samples. Hence, a batch of noise is input into the generator to obtain samples $\tilde{\mathbf{x}} = g_\theta(\mathbf{z})$. As the discriminator is a binary classifier, the cross-entropy loss (4.21)

$$\mathcal{L}_d = -\mathbb{E}_{\mathbf{x}\sim p_r}\left[\log(d_\varphi(\mathbf{x}))\right] - \mathbb{E}_{\mathbf{z}\sim p_z}\left[\log(1 - d_\varphi(g_\theta(\mathbf{z})))\right] \qquad (18.7)$$

is used as objective, where $\mathbb{E}$ denotes the expectation value, i.e. the average calculated over a batch. The classification labels 'real' or 'fake' are one-hot-encoded (4.19) by separating the expectation values in sampling from $\mathbf{x} \sim p_r$ or $\mathbf{z} \sim p_z$, respectively. The network predictions vary between $d_\varphi = 1$ for real samples and $d_\varphi = 0$ for clear fakes. By *minimizing* this loss, the classification between fake and real samples is learned. At the beginning of the adversarial training, the classification task is straightforward since the generator network produces only noise, but it becomes increasingly difficult as the generator improves.

Generator update: After training the discriminator, its feedback is used to train the generator. The basic concept of this update is that the generator is trained to fool the discriminator. Therefore, the output of the generator is used as input for the discriminator $d_\varphi(\tilde{\mathbf{x}}) \to d_\varphi(g_\theta(\mathbf{z}))$ so that the generator can directly obtain gradient-based feedback through the discriminator. Here φ, θ denote trainable parameters. To train the generator, the cross-entropy-like loss

$$\mathcal{L}_g = -\mathbb{E}_{\mathbf{z} \sim p_z}[\log(1 - d_\varphi(g_\theta(\mathbf{z})))], \quad \varphi = \text{fixed} \qquad (18.8)$$

is used.[4] By *maximizing* (18.8), the generator is trained to adapt the weights θ in a way, that the discriminator classifies the generated samples as real $(d_\varphi(g_\theta(\mathbf{z})) = 1)$. Note that this update directly incorporates knowledge of the discriminator to distinguish between generated and real samples. In other words, the generator performance is improved using the learned features of the discriminator, which describe the differences between real and generated samples. During the generator training, the weights φ of the discriminator are fixed and not updated. Otherwise, the update would result in a worse discriminator.

In summary, adversarial training can be formulated via

$$\mathcal{L}_{GAN} = -\mathbb{E}_{\mathbf{x} \sim p_r}[\log(d_\varphi(\mathbf{x}))] - \mathbb{E}_{\mathbf{z} \sim p_z}[\log(1 - d_\varphi(g_\theta(\mathbf{z})))], \qquad (18.9)$$

as a zero-sum game, where one player minimizes and the adversary maximizes the same loss, hence, minimizing and maximizing (18.9) results in a non-stationary optimization problem.[5] In other words, updating the generator will highly affect the next update of the discriminator and the other way round. Due to this dependency, it can be useful to reduce the momentum of various optimizers (see Sec. 4.8.3).

Further note that although the framework is designed to train the generator, the training's success relies mainly on the discriminator. By maximizing (18.8), the generator directly uses the discriminator feedback to improve its performance. This is a decisive difference to the objectives used for supervised training, where the attempts were made to find a local minimum in a stationary optimization problem. This major difference has the consequence that GANs are delicate to train and often show no 'convergence'.

---

[4]Note that the first term of the cross-entropy is missing as it vanishes when making the derivative w.r.t. the generator.
[5]It is further equivalent [181] to finding a Nash equilibrium.

Example 18.4. Generation of cosmic-ray footprints: In Fig. 18.6, a graphical sketch is shown illustrating the training of generative adversarial networks. The generator is trained to map from the latent space $\mathbf{z}$ to the space of cosmic-ray events $\mathbf{x}$ of an observatory (see Example 11.2). In the first epoch, the generated distribution p_θ (blue) does not match the real distribution p_r (black). During the training, the generator can approximate the real distribution better and better, resulting in an improved quality of the generated footprints (bottom). In the beginning, only noise is generated, then slowly first patterns are formed. Finally, realistic signal patterns are generated with meaningful signal amplitudes and the characteristic of a steep falling signal with distance to the cosmic-ray impact point.

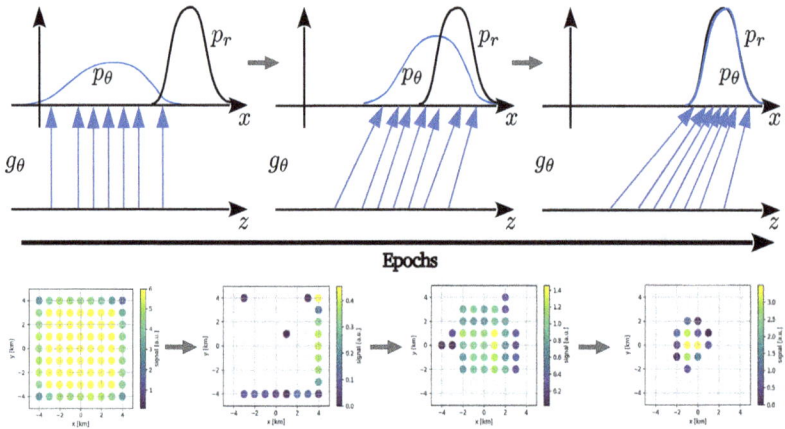

Fig. 18.6. Illustrating the training of generative adversarial networks.

18.2.2 *Conditioning*

If the generator should not only be able to generate new data, but data with certain properties, e.g. an image of a *spiral* galaxy or a particle shower with particular energy, the adversarial concept can be extended using *label conditioning* [182,183]. To achieve this, in addition to a noise vector $\mathbf{x} \sim p_z$, a vector $\mathbf{c} \sim p_c$ describing the desired properties is fed to the generator. These properties can be class labels (e.g. spiral or elliptical galaxies) or continuous properties like the energy of a particle shower. Hence, instead of learning the entire density at once, the generator is trained to learn a conditional density $\tilde{\mathbf{x}} \sim g_\theta(\mathbf{z}|\mathbf{c})$. For the conditioning of GANs, different approaches exist to condition the generation process. On the one hand, the discriminator can be trained to reconstruct the vector $\mathbf{c}$ describing the properties of data samples. On the other hand, the vector $\mathbf{c}$ can serve as additional input for the discriminator as well. In general, additional information is beneficial for deep networks. Hence label conditioning can help to stabilize the training and to improve the performance.

18.2.3   *Design of GANs*

In recent years several techniques were proposed which help to stabilize the training of GANs [184–187]. In the following, a short guideline is given which helps to design GANs with improved training stability:

- Use architectures which exploit the symmetries of your data (e.g. replace fully-connected layers with convolutional layers),
- Prevent sparse gradients (feedback) since extensive discriminator feedback is a key for successful training,
    - Do not use (max) pooling layers in the architecture, but convolutions with strides instead (compare with Chapter 8),
    - Use LeakyReLU activations in the discriminator instead of ReLUs to enable feedback for negative activations (Fig. 5.8),
- Make use of batch normalization in the generator and discriminator to stabilize training dynamics[6] (see Sec. 7.6).
- Make use of label conditioning (in generator and discriminator, see Sec. 18.2.2).

An example architecture is discussed in box 18.5.

---

[6]The most recent developments, however, do not focus on the explicit design of the framework, but on the fundamental training dynamics, e.g. by changing the objective. This is discussed further in Secs. 18.2.5 and18.2.6.

Example 18.5. Design of a generator and discriminator:

Generator: The generator network is trained to learn a mapping from the latent space into the space of real samples and thus has to meet several criteria. Since the generator's input consists of a noise vector and the output usually features an image-like data structure, using a reverse pyramidal topology in the generator helps to support a simple and structured latent space. Such a topology is illustrated in Fig. 18.7. The input latent vector is transformed using a fully-connected layer before being reshaped in an image-like structure. Transposed convolutions or rather upsampling followed by a standard convolution can be used to increase the image size (compare with Sec. 8.4). See Ref. [188] for details about checkerboard artifacts caused by transposed convolutions.

Another key ingredient for successful training is the activation function of the last layer. This activation function must always be able to cover the range of the data completely. For data to be generated which lies between $-1 \leq x \leq 1$, for example, the hyperbolic tangent would be a suitable activation.

Discriminator: The discriminator is used to map from the space of images to a vector and can therefore be implemented as a standard (convolutional) network suited to analyze the type of data at hand. For discriminators using cross-entropy like losses, a sigmoid (softmax) activation after the final layer has to be used:

$$\sigma(x) = \frac{1}{1 + e^{-x}} \tag{18.10}$$

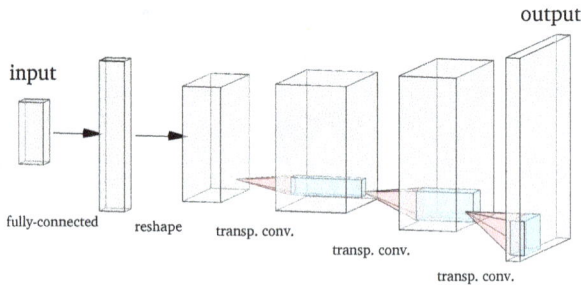

Fig. 18.7.   Example of a generator architecture featuring a pyramidal topology. Created using [19].

18.2.4 Challenges in the training of GANs

Besides the complex training of GANs, several more challenges arise, which are discussed in the following.

18.2.4.1 *Mode collapsing*

Mode collapsing, sometimes also called *helvetica scenario*, states that during the training of GANs, only particular samples can be generated. Therefore, the generator generates samples of a restricted phase space only. For example, training on galaxy data could result in a generator that can only generate samples similar to spiral galaxies. Especially the traditional set-up of GANs [180] strongly suffers from mode collapsing.

During the adversarial training, the generator slowly learns to generate samples that show characteristics similar to real samples. The observation can be explained as follows. In the beginning, only a small part of the real distribution is learned (see Fig. 18.6). Therefore, the generator will likely cover only a few modes of the true data. If the generator can produce samples of a particular mode, the discriminator will mostly focus on features related to this mode to identify the generated samples. It will also focus on distinct modes the generator does not cover to identify real samples. A discriminator focusing only on a restricted phase space will lead to a generator with a restricted phase space.[7] As the discriminator can detect the current mode of the generator, it will push the generator away from the current mode towards the next vulnerable mode, leading to a cyclic behavior. In simple terms, the mode collapsing phenomenon can be understood as the generator overtrains again and again on the discriminator.

18.2.4.2 *Vanishing gradients*

Learning a distribution is a complex task. At the beginning of the training, the discriminator will be able to separate generated (almost pure noise) and real samples with high confidence. Therefore, during the generator update the discriminator will output values of the sigmoid function (18.10) close to 0. In terms of the discriminator output (class score) f_θ directly before the sigmoid activation, this results in a large negative value $f_\theta \ll 0$.

Let us calculate the gradient update for the parameters θ of the generator $g_\theta(z)$ by using the binary cross-entropy loss (18.8) and considering the

---

[7]In fact, it happens that a few gradients become very large and 'dominate' the update pushing the generator to a single mode. This behavior is discussed in Sec. 18.2.5.

sigmoid activation function (18.10):

$$\frac{\partial \mathcal{L}_g}{\partial \theta} = \frac{\partial}{\partial \theta} \log(1 - d(g_\theta(z))) = \frac{\partial}{\partial \theta} \log\left(1 - \sigma(f_\theta(z))\right)$$

$$= \frac{\partial}{\partial \theta} \log\left(1 - \frac{1}{1 + e^{-f_\theta}}\right) = -\frac{e^{f_\theta}}{1 + e^{f_\theta}} \cdot \frac{\partial f_\theta}{\partial \theta}$$

The gradient, which should guide the generator, vanishes for a confident discriminator ($f_\theta \ll 0$). To avoid this situation, the discriminator cannot be trained to optimality. Its training needs to remain 'unfinished' during generator training. However, this will lead to noisy gradients of the discriminator, which will affect the quality of generated samples. Thus, a reasonable balance is to be found for the iterative training of the two networks.[8]

18.2.4.3 *Monitoring convergence*

In contrast to discriminative models, where evaluation metrics are generally easy to formulate, the performance evaluation of generative models is more complex. To estimate the quality of the generative model, the similarity between the target distribution p_r and the current approximation p_θ needs to be measured. During training, the performance of the generator is evaluated by a non-perfect discriminator to avoid vanishing gradients. Since high-dimensional distributions are approximated, several 'traditional' divergences are not suitable as the similarity measure. In fact, training using a perfect discriminator would minimize [180, 186, 190] the Jensen–Shannon (JS) divergence, a symmetrized version of the Kullback–Leibler (KL) divergence:

$$\mathcal{D}_{JS}(p_r||p_\theta) = \mathcal{D}_{KL}(p_r||p_m) + \mathcal{D}_{KL}(p_\theta||p_m) \qquad (18.11)$$

Here $\mathcal{D}_{KL}(p_r||p_\theta)$ is the Kullback–Leibler divergence (18.6) and the symmetrization is defined via $p_m = \frac{1}{2}(p_r + p_\theta)$. If p_θ should approximate p_r in a high-dimensional setting, it is most likely that during training the distributions are disjoint. For disjoint distributions, it is $\mathcal{D}_{JS}(p_r||p_\theta) = $ const. This implies that the loss is constant and does not take into account how close p_θ is to p_r. From this, it follows that the Jensen–Shannon divergence will not provide meaningful feedback for disjoint distributions.

In a figurative sense, think of the discriminator as a network that learns an embedding of the data and measures the similarity of the approximated

---

[8]To avoid vanishing gradients, several alternative loss functions (e.g. Ref. [180, 189]) were proposed. Still, they are not able to fundamentally change the training dynamics [186].

distribution p_θ and the real distribution p_r using a distance measure. For reasonable embedding, distributions in the direct neighborhood share similar features and are hence more similar than distributions that are far off. But if the loss is independent of the distance, it is not suitable as a proper similarity measure if the distributions are disjoint. But in fact, this is the case in GAN training. Already in the beginning, the discriminator can perfectly discriminate real and generated samples, meaning disjoint distributions. This further explains the observation that the image quality of generated pictures does not correlate with the discriminator loss [190].

18.2.5 *Wasserstein GANs*

Recently Wasserstein GANs (WGANs) [190, 191] were proposed to overcome vanishing gradients and the challenge of finding a suitable metric for disjoint distributions. The basic idea is to replace KL-based objectives with a softer metric — the *Wasserstein distance* — which provides a meaningful distance measure for disjoint distributions. The Wasserstein distance is also called Earth mover's distance, as it can be interpreted as the minimum amount of cost needed to transform one heap of earth into another [Fig. 18.8(a)]. In principle, the definition of cost is equal to the physics definition of work needed to move the earth. Hence, the cost is represented by the amount of dirt multiplied by the distance the earth has to be moved. This definition implies that this similarity metric provides a meaningful measure also for disjoint distribution as it directly scales with the distance of the distributions.

With the mathematical theorem of the Kantorovich–Rubinstein duality [192], the Wasserstein distance can be approximated using a discriminator. Thus, the differences of samples generated with a generator g_θ and real samples can be examined by a discriminator d_φ. The training loss for the training of WGANs reads:

$$\min_g \max_{d \in \mathrm{Lip}_1} \{\mathbb{E}_{\mathbf{x} \sim p_r}[d_\varphi(\mathbf{x})] - \mathbb{E}_{\mathbf{z} \sim p_z}[d_\varphi(g_\theta(\mathbf{z}))]\} \qquad (18.12)$$

The adversarial training using two networks is similar to the basic setup described in the previous sections. However, the decisive difference is that for the discriminator, the binary classification task (18.7) is replaced here by a distance measure (18.12).

But here, Lip_1 denotes that the discriminator is required to be a 1-Lipschitz function meaning, that its adaptive parameters φ are confined by a Lipschitz constraint. A 1-Lipschitz constraint means that the network has to have a derivative smaller than $|1|$ everywhere. To estimate the Wasserstein distance, i.e. training the discriminator, the difference between the discriminator output when applied on the fake and the real samples is maximized. The Lipschitz condition ensures that this expression does not diverge. Hence, the discriminator can be trained till convergence which assures a precise estimate of the Wasserstein distance. Since, in general, a neural network can approximate any arbitrary function, this condition must be explicitly enforced.

The *gradient penalty* [191] can be used to implement the Lipschitz condition. By extending the loss (18.12) by a second term

$$\mathcal{L}_{GP} = \lambda \, \mathbb{E}_{\hat{\mathbf{u}} \sim p_{\hat{\mathbf{u}}}} [(\| \nabla_{\hat{u}} d_\varphi(\hat{\mathbf{u}}) \|_2 - 1)^2] \qquad (18.13)$$

gradients unequal to norm 1 are punished.[9] Here, λ is a hyperparameter (usually set to 10), and

$$\hat{\mathbf{u}} = \epsilon \mathbf{x} + (1 - \epsilon) \tilde{\mathbf{x}},$$

are points which are pairwise and randomly sampled ($0 \leq \epsilon \leq 1$) on straight lines between generated $\tilde{\mathbf{x}}$ and real samples $\mathbf{x}$. Constraining the gradient allows to approximate the Wasserstein distance using the discriminator and ensures adequate gradients even outside the support of p_r and p_θ. A visualized description of the gradient penalty is shown in Fig. 18.8(b).

Even if changing the objective heavily influences the internal training dynamics, the basic design of GANs (Example 18.5) can be adapted for WGANs[10] by removing the sigmoid in the discriminator.

When employing the Wasserstein distance, the training of the adversarial framework is changed in such a way that the discriminator can be trained to convergence. Usually, around ten discriminator updates should be followed by a single generator update. As the Wasserstein distance is suitable for disjoint distributions, the measure will not diverge and give a reasonable similarity measure that correlates with the quality of generated samples. The discriminator further delivers adequate gradients as they are explicitly constrained by the gradient penalty. Additionally, in contrast to the traditional setup, no sigmoid function is needed as the last activation function in the discriminator. This removes vanishing gradients as the

---

[9]This is not equivalent to the Lipschitz constraint $-1 \leq m \leq 1$, but tests [191] show slightly better performance by constraining the gradient norm to exactly 1.

[10]Consider to use layer normalization [193] instead of batch normalization in the discriminator because the gradient penalty is calculated between sample pairs. Furthermore, activations with non-continuous second-order derivatives (e.g. ELU [34]) cannot be used, as they induce interference terms in the gradient penalty.

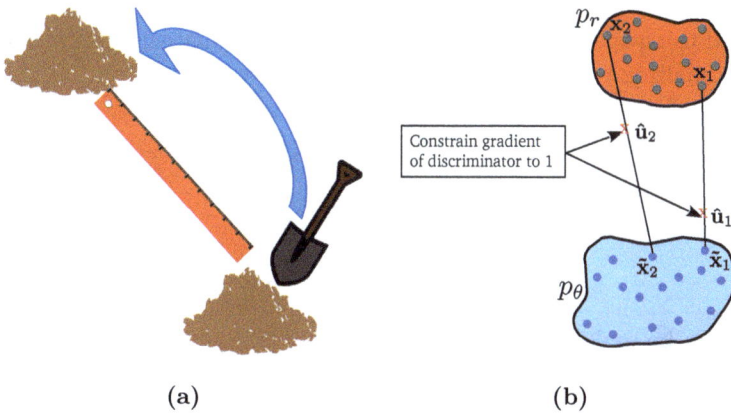

(a)   (b)

Fig. 18.8. (a) Illustration of the Wasserstein metric as Earth mover's distance. The Wasserstein distance amounts to the minimum cost (mass × distance) needed to transform one heap of earth into another. (b) Illustration of the gradient penalty. The gradient of the discriminator evaluated at the events mixtures $\hat{u}_1, \hat{u}_2$, created using random sampling, is constrained to 1. This leads to normalized gradients even outside p_θ and p_r.

objective scales linearly with the distance. Due to the mentioned improvements, the discriminator is often called a *critic* as now the focus changes from simple discrimination to adequate feedback.

18.2.6 *Spectrally normalized GANs*

Investigating the success of WGANs from a different perspective [194] introduced *spectrally normalized* GANs (SNGANs). The basic idea behind SNGANs is that the training dynamic of GANs is highly dependent on the gradients of the discriminator. Hence, the success of WGANs is not only related to the approximation of the Wasserstein distance but also because the gradients are properly constrained. Put simply, during the training, the discriminator receives only samples from the real distribution and generated samples from the current state of the generator. In the course of the training, samples that lie somewhere between the distributions will occur most likely. The gradients with respect to such samples are noisy and can become very large [186]. Therefore, regularization of these gradients is important and implicitly used in WGANs using the gradient penalty [see Fig. 18.8(b)].

In SNGANs, the Lipschitz constraint, i.e. the gradient regularization, is replaced using spectral normalization of the discriminator[11] weight matrix. The normalized weight matrix $\mathbf{W}_{\text{norm}}$ is given by

$$\mathbf{W}_{\text{norm}} = \frac{\mathbf{W}}{\sigma(\mathbf{W})}, \tag{18.14}$$

where $\sigma(\mathbf{W})$ denotes the spectral norm of the weight matrix. The spectral norm is defined via

$$\sigma(\mathbf{W}) = \max_{x \neq 0} \frac{||\mathbf{W}\mathbf{x}||_2}{||\mathbf{x}||_2}, \tag{18.15}$$

which can be understood as maximum factor a vector $\mathbf{x}$ can be stretched after a multiplication with $\mathbf{W}$. Estimating the spectral norm can be computationally expensive, therefore, the power iteration method [197, 198] is recursively applied[12] since the norm is expected to change not rapidly.

The principle of SNGANs can be understood as follows. Each layer of the discriminator gets as input a feature vector $\mathbf{y}$ from the layer before. This vector is rotated and stretched in different ways by multiplying it by the weight matrix $(\mathbf{Wy})$. An extensive stretching in one direction would be equal to a large sensitivity to a specific feature or rather a gradient that would dominate the update. As the spectral norm of a matrix is the maximum possible stretching factor of a unit vector when being transformed using the matrix, normalizing the weight matrix directly counteracts such extensive stretching.

This perspective makes it especially clear why traditional GANs are especially prone to mode collapse. Since the gradients are not normalized in the discriminator, the framework is susceptible to a few strong gradients. These dominate the update, and the generator collapses into a single mode.

To further prevent vanishing gradients, usually the alternative generator loss [180]

$$\mathcal{L}_g = -\mathbb{E}_{\mathbf{z} \sim p_z} \log(d_\varphi(g_\theta(\mathbf{z}))) \tag{18.16}$$

is *minimized* in SNGANs (compare with (18.8)). Altogether the formulation of SNGANs has the advantage that a one-by-one update of generator and discriminator is adequate. This heavily reduces computational costs.

In principle, it is even possible to combine WGANs and spectral normalization in an adversarial framework because both methods have the same goal but are not equivalent and feature different advantages [194].

---

[11] In principle, also the application of spectral normalization in the generator can improve the performance as well [195, 196].

[12] See Appendix of [194].

18.2.7 Additional advances

Currently, the field of generative models evolves relatively fast, and many different aspects are subject to ongoing research [199]. Additionally, the success of generating high-quality samples strongly depends on the data to be generated. Hence, one should be aware that not all advances on typical computer science data sets are always transferable one-to-one to physics data sets. Usable are, of course, theoretically founded improvements.

Whereas the training dynamic mainly depends on the discriminator network, the quality of the generated samples also depends on the generator itself. Also, here recent progress was made. Again it is crucial to stress that fundamental knowledge about the data is of great importance. Using suitable transformations (e.g. convolutions) will always help to improve performances.

For the creation of large images, for example, the concept of attention [200] is used to achieve global coherence [196, 201]. Furthermore, the decomposition of images into low-, mid-, and high-level features [202, 203] have led to exceptionally impressive examples[13] in the generation of human faces.

18.3 Normalizing flows

Finally, let us briefly highlight a third technique for implementing generative models: so-called normalizing flows [204, 205]. Like GANs and VAEs, such flows aim to learn the underlying probability distribution of an observed data sample. However, in addition to generating new samples, flows also allow a direct evaluation of the probability density.

Invertible mapping: The idea behind flows is to apply a sequence of invertible and differentiable mappings to a simple probability distribution (e.g. normal distribution) in order to reshape it to approximate the observed complex distribution. This is illustrated in Fig. 18.9. To generate new samples, one can then sample from the simple distribution and apply the transformations to emulate the observed distribution. In this respect, flows can be thought of as a VAE where the decoder is enforced to be the inverse of the encoder, and one has complete control over the shape of the latent-space distribution. Another difference is that GANs and VAEs usually operate with a latent space that has lower dimensionality than the data

---

[13]Visit https://www.whichfaceisreal.com to see some of the striking examples.

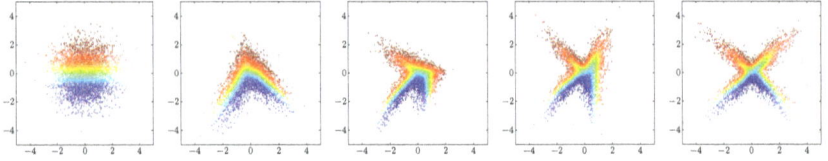

Fig. 18.9. Example of a flow which transforms samples from a normal distributed base density (left-most image) to a cross-shaped density in four steps [207].

while flows — due to the invertible nature of the transformation — have a latent space of equal dimensionality [206].

For learning a density, one starts from an observed data point and uses the *inverse transformations* until arriving at the sample probability of the simple distribution. The density of this data point can then be calculated as the product of the changes in volume from the inverse transformations, i.e. the Jacobian determinant of each mapping and the density in the tractable simple probability distribution. The ability to directly obtain a value for the density at a given point is an advantage of flow-based models over GANs or VAEs. We do not need this property for our generative task at hand but will return to it when discussing anomaly detection in Chapter 20.

Formally, we have data points in M dimensions $\mathbf{x} \in \mathbb{R}^M$ originating from the probability distribution $p_{\text{data}}(\mathbf{x})$ as well as the latent space following the simple distribution $p(\mathbf{z})$. The generator function g allows generating new samples by using $g(\mathbf{z})$ for $\mathbf{z} \sim p(\mathbf{z})$.

For the learning process, we start from the inverse function $f = g^{-1}$. Thus, we describe here the mapping between the two probability distributions using the function f. For the transformation of a one-dimensional probability distribution $p(x)$ to the distribution $p(z)$ we conserve probability by $p(x)\,dx = p(z)\,dz$. For mapping the M-dimensional distribution, we need the Jacobian determinant:

$$p(\mathbf{x}) = p(\mathbf{z}) \cdot \left| \frac{\partial \mathbf{z}}{\partial \mathbf{x}} \right| \tag{18.17}$$

Since the latent variable results from $\mathbf{z} = f(\mathbf{x})$, it is $p(\mathbf{z}) = p(f(\mathbf{x}))$. The desired probability distribution $p(\mathbf{x})$ can thus be evaluated directly from the function f

$$p(\mathbf{x}) = p(f(\mathbf{x})) \cdot |\det \mathbf{J}_f(\mathbf{x})| \tag{18.18}$$

using the determinant of the Jacobian $|\det \mathbf{J}_f(\mathbf{x})|$ for the transformation from $\mathbf{z}$ to $\mathbf{x}$.

So far, we have not discussed how such a model can be constructed, i.e. how a function g with suitable properties can be found. The first important observation is that a composition of N bijective functions $g = g_N \circ g_{N-1} \circ \ldots g_1$ is invertible with

$$f = g_1^{-1} \circ \ldots \circ g_{N-1}^{-1} \circ g_N^{-1} \qquad (18.19)$$

$$= f_1 \circ \ldots \circ f_{N-1} \circ f_N. \qquad (18.20)$$

The Jacobian determinant of the composition of functions is the product of the individual determinants:

$$|\det \mathbf{J}_f| = \prod_{i=1}^{N} |\det \mathbf{J}_{f_i}| \qquad (18.21)$$

This means that we can have a sequence of relatively simple nonlinear bijective functions and transform the initial probability step-by-step into the desired shape.

Invertible building blocks: A large and rapidly developing number of variants exists for these building blocks [205, 207]. Let us now review in detail how networks can be made invertible with a simple-to-calculate Jacobian determinant. A concrete realization of an invertible network block transforming $\mathbf{z} = f(\mathbf{x})$ is shown in Fig. 18.10 [209]. In the *forward pass*, the input data $\mathbf{x} \in \mathbb{R}^M$ is split into two parts $\mathbf{x}_1 = \mathbf{x}_{1:M/2}$ and $\mathbf{x}_2 = \mathbf{x}_{M/2+1:M}$. This is followed by symmetric processing of the two parts. First, component $\mathbf{x}_2$ is mapped by neural network s_2 and subjected to the exponential function before being multiplied element-wise by component $\mathbf{x}_1$. The component $\mathbf{x}_2$ is mapped by another network t_2 whose result is added to it element by element. The result is denoted as $\mathbf{z}_1$. Subsequently, $\mathbf{z}_1$ is now merged element-wise with $\mathbf{x}_2$ via two networks s_1 and t_1. Concatenation $\mathbf{z} = (\mathbf{z}_1, \mathbf{z}_2)$ yields the output of the network block.

The *backward pass* operates with reversed signs. The division can be achieved by element-wise multiplication by $\exp(-s_i)$ where $i = 1, 2$. While the networks s_i and t_i themselves are not invertible — and do not need to be, as they are always used in forward mode — the overall block which maps between $\mathbf{x}$ and $\mathbf{z}$ is invertible.

For the invertible block to be useful in practice, we also need to calculate the determinant of the Jacobian. We can view the forward pass as two subsequent transformations f_1 and f_2 — corresponding to the left and right halves of Fig. 18.10 respectively — applying the following changes to

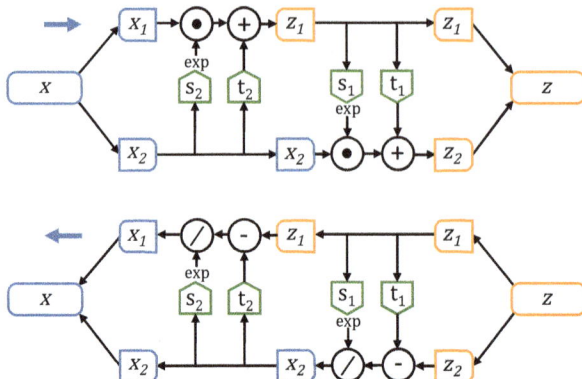

Fig. 18.10. Example for an invertible mapping using a real-valued non-volume preserving (real NVP) transformation [208]. Here, s_i and t_i ($i = 1, 2$) denote networks. The upper diagram gives the forward pass $x \to z$ for training the network. The lower diagram shows the backward pass $z \to x$ enabling generation of distributions in x from the normally distributed z.

the data:

$$\begin{pmatrix} \mathbf{x}_1 \\ \mathbf{x}_2 \end{pmatrix} \xrightarrow{f_1} \begin{pmatrix} \mathbf{z}_1 \\ \mathbf{x}_2 \end{pmatrix} \xrightarrow{f_2} \begin{pmatrix} \mathbf{z}_1 \\ \mathbf{z}_2 \end{pmatrix} \tag{18.22}$$

As the structure for f_1 and f_2 is similar, we first focus on f_1:

$$\mathbf{x}_1 \xrightarrow{f_1} \mathbf{z}_1 = \mathbf{x}_1 \odot \exp(s_2(\mathbf{x}_2)) + t_2(\mathbf{x}_2) \tag{18.23}$$

$$\mathbf{x}_2 \xrightarrow{f_1} \mathbf{x}_2. \tag{18.24}$$

The Jacobian matrix for this transformation $\mathbf{J}_1$ reads:

$$\mathbf{J}_1 = \begin{pmatrix} \frac{\partial \mathbf{z}_1}{\partial \mathbf{x}_1} & \frac{\partial \mathbf{z}_1}{\partial \mathbf{x}_2} \\ \frac{\partial \mathbf{x}_2}{\partial \mathbf{x}_1} & \frac{\partial \mathbf{x}_2}{\partial \mathbf{x}_2} \end{pmatrix} \tag{18.25}$$

$$= \begin{pmatrix} \mathrm{diag}(\exp(s_2(\mathbf{x}_2))) & \frac{\partial \mathbf{z}_1}{\partial \mathbf{x}_2} \\ 0 & \mathbb{1} \end{pmatrix}. \tag{18.26}$$

By construction, we arrived at a triangular matrix. This shape greatly simplifies the calculation of the determinant:

$$\det \mathbf{J}_1 = \prod \exp(s_2(\mathbf{x}_2)) = \exp\left(\sum s_2(\mathbf{x}_2)\right). \tag{18.27}$$

Here, the sum goes over the output dimension of s_2. In the same way, the Jacobian determinant for the second half of the transformation f_2 can be calculated to be

$$\det \mathbf{J}_2 = \exp\left(\sum s_1(\mathbf{z}_1)\right). \tag{18.28}$$

Combining these shows the simple form of the overall determinant of the forward pass:

$$|\det \mathbf{J}_f| = \exp \left(\sum s_2(\mathbf{x}_2) + \sum s_1(\mathbf{z}_1) \right) = \exp \left(\sum s(\mathbf{x}) \right). \qquad (18.29)$$

For the last equality, we simplified the notation to highlight that the determinant is the exponential function applied to a sum of network predictions s. When multiple such blocks are applied in sequence, due to (18.21), we just gain additional terms in that sum.

To summarize, by splitting the input features into two parts, we notice how a transformation block, that is invertible and allows calculating the change in probability volume, can be constructed from standard (i.e. non-invertible networks) and basic mathematical operations. When more expressiveness is needed, multiple such blocks can be applied subsequently. An alternative construction based on autoregressive transformations is sketched in Example 18.6.

Example 18.6. Autoregressive flows: A popular alternative building block for invertible networks are *masked autoregressive flows* (MAFs) [210]. An autoregressive flow is a bijective function of a number of inputs $\mathbf{y}_t$ which for each output $\mathbf{x}_t$ is conditioned on all preceding inputs:

$$\mathbf{x}_1 = h(\mathbf{y}_1, \theta_1) \qquad (18.30)$$

$$\mathbf{x}_2 = h(\mathbf{y}_2, \theta_2(\mathbf{x}_1)) \qquad (18.31)$$

$$\mathbf{x}_3 = h(\mathbf{y}_3, \theta_3(\mathbf{x}_1, \mathbf{x}_2)) \qquad (18.32)$$

$$\ldots \mathbf{x}_t = h(\mathbf{y}_t, \theta_t(\mathbf{y}_1, \ldots, \mathbf{y}_{t-1})) \qquad (18.33)$$

where h is a bijective function parameterized by θ. This construction again leads to a triangular form of the Jacobian determinant and therefore to a simple and computationally efficient calculation of the determinant.

Generator training: Next, let us consider how such a model can be applied and trained in practice. When used as generator, we start with the distribution $p(\mathbf{z})$ of the latent variable $\mathbf{z}$ and should produce distributions $g = f^{-1} = p_{\text{model}}(\mathbf{z}, \theta)$ that approximate the data distribution $p_{\text{data}}(\mathbf{x})$. In Fig. 18.10, generation corresponds to running in *backward mode* (lower figure).

To find model parameters θ (i.e. weights of the component networks) that best approximate p_{data}, the objective function needs to minimize the negative log-likelihood of the training samples[14] (compare with Sec. 14.2):

$$\mathcal{L} = -\mathbb{E}_{\mathbf{x} \sim p_{\text{data}}} \left[\ln p_{\text{model}}(\mathbf{x}, \theta) \right]. \tag{18.34}$$

For the network training, we use the *forward pass* (upper part of Fig. 18.10) and input data $\mathbf{x}$ from p_{data}. For the mapping of the distribution $p_{\text{data}}(\mathbf{x})$ onto the distribution $p(\mathbf{z})$ we conserve probability by taking into account the Jacobian determinant already calculated above (18.18):

$$\mathcal{L} = -\mathbb{E}_{\mathbf{x} \sim p_{\text{data}}} \left[\ln \left(p(f(\mathbf{x})) \left| \det \mathbf{J}_f \right| \right) \right]$$

Any reasonable choice of $p(\mathbf{z})$ is possible. However, due to the logarithm, a standard normal distribution is especially convenient as in one dimension it is:

$$\ln p(z) = \ln \left(e^{-z^2/2} \right) = -\frac{(f(x))^2}{2} \tag{18.35}$$

For high-dimensional distributions we use the Euclidean squared distance $\|f(\mathbf{x}))\|_2^2$. Also inserting the explicit form of the Jacobian determinant (18.29) yields:

$$\mathcal{L} = -\mathbb{E}_{\mathbf{x} \sim p_{\text{data}}} \left[-\frac{1}{2} \|f(\mathbf{x}))\|_2^2 + \sum s(\mathbf{x}) \right]$$

Using the batch size k, the objective function to be minimized finally becomes

$$\mathcal{L} = \frac{1}{k} \sum_{i=1}^{k} \left(\frac{1}{2} \|f(\mathbf{x}_i))\|_2^2 - \sum s(\mathbf{x}_i) \right). \tag{18.36}$$

In this way, the network in the lower Fig. 18.10 will generate new data $\mathbf{x}$ from Gaussian distributed latent variables $\mathbf{z}$ which approximate $p_{\text{data}}(\mathbf{x})$. The precise form will be different for other implementations of normalizing flows, but the underlying idea of building bijective mappings that allow tracking the change in probability volume remains.

Compared to VAE and GAN, normalizing flows are a relatively recent — yet promising — addition to the repertoire of generative models [199]. We highlight one application in describing the fundamental interactions of hadrons in Example 18.7. Other applications in physics [213–215] exist as well, and we will also encounter flows again when unfolding data distributions in Chapter 19 and searching for anomalies in Chapter 20.

---

[14]This is an important difference to the earlier introduced VAEs: There, only an approximation — specifically a lower bound on the likelihood — could be obtained [176]. Flows, due to the tractable nature of the likelihood calculation, allow optimizing the likelihood directly.

However, for a discussion of other practical applications of generative models, we focus on more established methods in the following section.

Example 18.7. Flows in lattice QCD: The theory of Quantum Chromodynamics (QCD) describes the so-called strong interaction of fundamental constituents of matter. Its coupling strength is inversely proportional to the energy, leading to asymptotic freedom at high energies and to strongly coupled theories at low energies. In this strong coupling limit, the theory becomes non-perturbative, meaning that an expansion in powers of the coupling strength will, in general, not converge. To circumvent this problem, calculations are carried out on a discretized spacetime lattice, often using Monte Carlo methods. However, these become increasingly inefficient in some phase space regions. Learning the probability distribution of physical observables and sampling from it — using generative models — is a promising alternative. A popular solution consists of training normalizing flow models for this problem. We point to Ref. [211] for a hands-on pedagogical introduction of flows in the context of lattice field theory.

A particularly interesting aspect is including symmetries of the underlying theory in constructing the flow model. In Ref. [212], the authors consider gauge transformations and show how the building blocks of normalizing flows (the coupling layers) can be made equivariant under certain symmetries — i.e. constructed in such a way that applying the symmetry commutes with the coupling layer.

18.4   Physics applications

18.4.1   *Performance evaluation*

For generative models, it is crucial to define metrics that can evaluate the quality of the generated samples. In computer science, mostly high-dimensional embeddings [181, 216] are used to obtain a data-driven performance estimate. In contrast, in physics, it is often essential that the known physics laws also apply to the generated samples. Furthermore, the different modes in a dataset are often strongly unbalanced, making it especially important to measure the sample diversity. This can make it necessary to define proper metrics that evaluate the generated samples or investigate certain phase-space regions in-depth, especially if an important mode is underrepresented in the training data.

18.4.2 *Example: generation of calorimeter images*

Deep generative models are particularly suitable for the generation of high-dimensional data. A physics example is energy depositions in a calorimeter detector (Fig. 18.11). In many particle detectors, calorimeters are used to measure the energy of particles traversing the detector. When a particle passes through the calorimeter, it interacts with the detector material and deposits energy. To obtain precise detector simulations, sophisticated simulation frameworks must be used to simulate the interaction between particles and matter and the detector to be read out. This is time-consuming and requires substantial computing resources. Generative models can help to increase the statistics of simulated events [217] or generate samples using models trained on test-beam data [218].

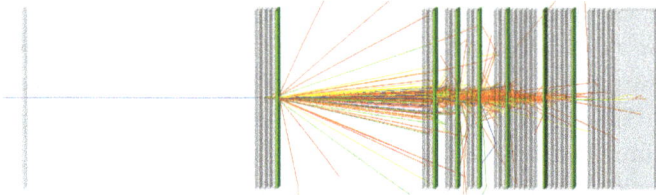

Fig. 18.11. Simulated calorimeter shower of an electron arriving from the left, producing numerous shower particles in the material. The seven green vertical lines represent sideways projections of the sensor planes recording the energy depositions, with their top views shown in Fig. 18.12(a) below. The gray areas denote passive absorber material.

Since most detectors feature approximately a homogeneous arrangement of calorimeter cells, image-like formats can be used which are perfectly suited for an analysis using convolutions. Note that several things changed after switching the dataset from natural images to physics. Especially the logarithmic distribution of signals and the sparsity offer a new challenge. A detailed description is given in Example 18.8. The upcoming upgrade of the large hadron collider (LHC) at CERN will significantly increase the luminosity of the accelerator. Hence much more simulations are required. Therefore, several collider experiments are currently investigating the acceleration of their detector simulations using generative models [219–225]. Besides, other works are focused on simulations using generative models, including cosmology [226] and astrophysics [227].

In Fig. 18.12, the generation of particle showers at a future multi-layer calorimeter [228] is shown. Besides evaluating the mean occupancy and the study of individual samples, the distributions of typical shower observables have been examined and compared to the simulation [229]. These examined observables were not constrained during the training but showed a good agreement with the simulation. Even interpolations of particle energies worked successfully. For showers initiated by particles with energies that were not part of the training set, the generated distribution (grey diamonds) matches the simulation (grey histogram).

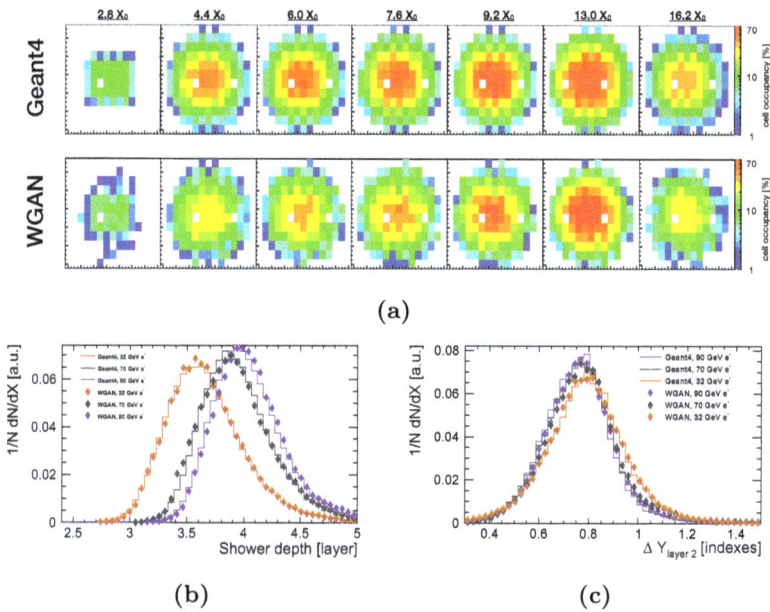

Fig. 18.12. Study of a WGAN used to generate signal patterns in a calorimeter. (a) Generated (WGAN) and simulated (GEANT4) pixel occupancies for 90 GeV electrons. Comparison between calorimeter observables computed using generated showers (symbols) and simulated showers (histograms) for the (b) shower depth and (c) the transverse shower spread. The 70 GeV showers (grey) were not part of the training. From [219] (©*Springer Nature Switzerland AG, reprinted by permission of Springer Nature*).

Example 18.8. Generating calorimeter data: The data structure of calorimeter data is image-like but exhibits decisive differences from natural images. It is of particular importance to recognize these differences to design the architecture of the generative model accordingly. Adjustments could be, e.g. the use of certain activations (ReLU as the last activation in the generator), the use of locally-connected convolutions (to set a prior on sparsity), or specific normalizations. An example comparison between natural images and calorimeter images is given below.

type	calorimeter data	natural images
Channels	number of layers	3 (RGB)
Resolution	relatively low	high resolution
Pixel distribution	sparse	coherent
Pixel range	several magnitudes	$0 - 255$
Intensity distribution	exponential	$\approx$ Gaussian
Correlations	mostly local	local, global

Example 18.9. String theory: Another potential application of generative models comes from string theory [230]. There, a so-called landscape of vacua describes possible realizations of the theory. As the number of such possible vacua is vast, it is necessary to analyze them statistically. Specifically, one wants to calculate the expectation value of physical observables given the probability distribution of vacua. Practically, for a subset of vacua, a Lagrangian description related to those observables exists. Even better, these descriptions can be represented by a simple 2D matrix — i.e. a grayscale image.

A generative model trained to reproduce such images corresponding to different vacua then allows efficient sampling and enables the calculation of observables. This is useful as classical calculations for this task are often too computationally complex. The authors of Ref. [230] also show that conditional generative models can learn to extrapolate for such predictions. To this end, a generative model was only trained on a subset of vacua — chosen to correspond to given values of a topological invariant characterizing such spaces — and then evaluated in simulating other values of the topological invariant.

18.5 Exercises

Exercise 18.1. Generation of fashion MNIST samples: Start from
Exercise 18.1 from the exercise page to develop Generative Adversarial
Networks for the creation of fashion images.

(1) Pre-train the discriminator and evaluate the test accuracy.
(2) Set up the main loop for training the adversarial framework:

- Save the generator and discriminator loss for every training
  step.
- Use uniform distributed noise as prior distribution for the
  generator.

(3) Generate some images after each training epoch. Plot the training losses for the generator and discriminator and generated images.

Exercise 18.2. Wasserstein GAN: Recall the description of a cosmic-
ray observatory in Example 11.2 and Fig. 11.2(b). In response to a
cosmic-ray-induced air shower, a small part of the detector stations is
traversed by shower particles leading to characteristic arrival patterns
(dubbed *footprints*, see Fig. 18.13). The number of triggered stations
increases with the cosmic-ray energy. The signal response is largest close
to the center of the shower.

Start from Exercise 18.2 from the exercise page to develop a generator
for air-shower footprints using a Wasserstein GAN:

(1) Build a generator and a critic network which allows for the generation of 9×9 air shower footprints.
(2) Set up the training by implementing the Wasserstein loss using
the Kantorovich–Rubinstein duality. Pretrain the critic for one
epoch and hand in the loss curve.
(3) Implement the main loop and train the framework for 15 epochs.
Check the plots of the critic loss and generated air shower footprints.
(4) Name four general challenges of training adversarial frameworks.
(5) Explain why approximating the Wasserstein distance in the dis-
criminator/critic helps to reduce mode collapsing.

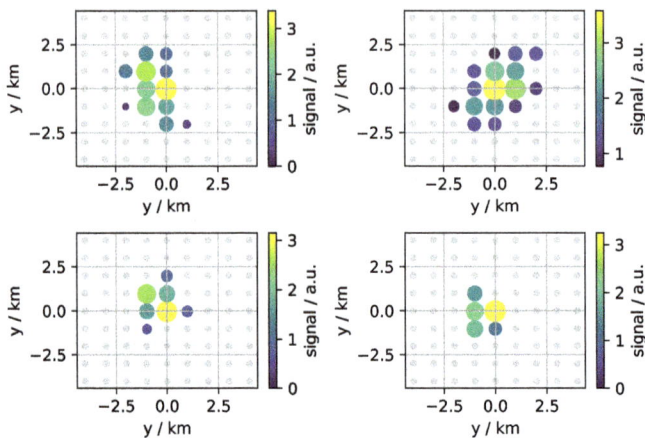

Fig. 18.13. Arrival patterns of cosmic-ray induced air showers, generated by a WGAN.

Summary 18.1.

(1) Generative models are trained to learn high-dimensional data distributions.

(2) By sampling from this learned distribution, new samples can be generated cost-efficiently.

(3) The performance of generative models is challenging to measure. It requires precise examinations of the complete phase space and detailed considerations of individual samples.

(4) Variational autoencoders (VAE) are a method related to autoencoders. They are based on an encoder learning a low-dimensional data encoding and a decoder that transforms the encoding back to the input space.

- During training, the learned encoding is constrained to match a specific distribution (usually a multivariate Gaussian), and further, the decoded sample must match the prior sample.

- After training, new samples can be obtained using the decoder by sampling from the multivariate (Gaussian) distribution which describes the data encoding.

(5) In generative adversarial networks (GAN), a hand-coded loss is replaced by a discriminator network evaluating the performance of a generator network. Figuratively, this adversarial training can be interpreted as a game where the generator network tries to fool the discriminator, which learns to identify real and generated samples. Mathematically it can be understood as divergence minimization.

- Due to the simultaneous training of two networks, GANs are delicate to train and require dedicated strategies and architectures.

- Using advanced techniques, e.g. approximating the Wasserstein distance with the discriminator or applying spectral normalization of the weight matrices, the training of GANs can be stabilized, and the performance significantly improved.

(6) Normalizing flows learn a bijective mapping from the distribution of data points to a latent space of predefined probability distributions — typically Gaussian. These transformations are constructed to be invertible and to allow a convenient calculation of the Jacobian. This allows two uses:

- Applying the transformation from latent space to data space allows sampling and acts as a generative model.
- The reverse transformation associates a density with each data point and can be used for density estimation (see e.g. Chapter 20).

Chapter 19

Domain adaptation, refinement, unfolding

Scope 19.1. A principal challenge for real-world applications of machine learning in physics comes from systematic differences between available training data and data in the application region. Solving this problem is known as *domain adaptation*. In this chapter, we will introduce several approaches for domain adaptation and related problems.

We already encountered several examples of insufficiently defined labels in Chapter 16. Here the problem is different: our training data can be assumed to be labeled fully and correctly; however, the training data noticeably differs from our application data. Taking into account and mitigating these differences is the goal of *domain adaptation.*

A prime example is a gap between simulation and reality: We can use simulations (e.g. Monte-Carlo-based or deep generative models — see Chapter 18) to create synthetic data of rare or unseen processes. Assuming a decent simulation quality, these samples will mimic the true (experimental) data well, but not perfectly. The need for domain adaptation can also arise when historical data is used to train the algorithm but is no longer accurate due to changes over time.

Formally, a domain $\mathcal{D}$ is a combination of a feature space and a probability distribution of data elements $p(\mathbf{x})$:

$$\mathcal{D} = \{\mathcal{X}, p(\mathbf{x})\} \tag{19.1}$$

The feature space $\mathcal{X}$ can be, e.g. the N-dimensional real number space $\mathcal{X} = \mathbb{R}^N$ which contains data elements $\mathbf{x} \in \mathcal{X}$. An n-dimensional example is $x = \{x_1, \ldots, x_n\} \in \mathcal{X}$.

Domain adaptation is needed when we train an algorithm on a source domain $\mathcal{D}_{\text{source}} = \{\mathcal{X}_{\text{source}}, p(\mathbf{x})_{\text{source}}\}$ but apply it on a different target domain $\mathcal{D}_{\text{target}} = \{\mathcal{X}_{\text{source}}, p(\mathbf{x})_{\text{target}}\}$.[1] We further assume that labels y_i^{source} are available for the data $\mathbf{x}_i$ in the source domain, while for the data in the target domain, no or only weak labels exist.[2]

If the distributions p_{source} and p_{target} were identical, no domain adaptation would be needed. If they differ, however, a machine learning algorithm trained on p_{source} will in general not make reliable predictions on p_{target}. This motivates *refinement* which is discussed in Sec. 19.1. Refinement consists in applying a transformation f to the source distribution so that it agrees with the target: $f(p_{\text{source}}) = p_{\text{target}}$. Algorithms can then be trained on $\mathbf{x} \sim f(p_{\text{source}})$ without additional modifications.

Alternatively, we can use p_{source} for training and instead transform distributions after a selection on the classifier output, for example, by using statistical weights. Such weights are termed *scale factors* and are introduced in Sec. 19.2.

Instead of changing the inputs (as in refinement) or the output (as when using scale factors) it is also possible to include stable predictions between the source and target domains as an additional objective in the loss function. Adversarial training can be used to ensure that only properties common between source and target are learned. We will explore this approach in Sec. 19.3.

Finally, we will consider the closely related problem of *unfolding* in Sec. 19.4. This means asking the question of how to *undo* stochastic distortions originating, e.g. from detector effects, and infer an underlying true distribution.

19.1   Refinement for simulations to appear data-like

Here we follow the idea that measurement data correspond to samples of a probability density distribution p underlying an experiment. Our initial goal is to quantitatively assess the differences between the measurement data of two experiments with probability densities p_1 and p_2. An example use case is an experiment with density p_1 whose simulated experiment follows density p_2. If $p_1 \neq p_2$, measured data and simulated data are different.

---

[1]Note that domain adaptation is a special case of *transfer learning* where not only the domain but also the learning task (e.g. regression or classification) might change. Transfer learning is not discussed in this work, but we refer the reader to [231] for an overview.

[2]Otherwise, we could directly train on $\mathcal{D}_{\text{target}}$ without the need for domain adaptation.

But, we assume the label distributions of p_1 and p_2 to be approximately similar.

Quantifying differences between experiments: Since there is no one-to-one mapping between the data from both experiments, we use an adversarial framework utilizing the Wasserstein distance, which we already introduced in Sec. 18.2.5. This distance measure quantifies the differences between the probability densities p_1 and p_2 given data samples from both experiments.

Figure 19.1(a) shows a possible adversarial refiner architecture which is related to GANs (Fig. 18.5). But the decisive difference is the input: instead of random numbers as used in GANs, data $\mathbf{x} \sim p_1$ from experiment 1 (here a simulation) is input to the first network (*Refiner*). The task of this network f is to modify the data from p_1 (source domain) to follow a distribution $\mathbf{x}' \sim p_1'$, where p_1' is indistinguishable from the data distribution p_2 (target domain) of the second experiment:

$$p_1 \neq p_2 : \quad \text{refinement} \quad p_1 \to p_1' = f(p_1) \quad \text{to obtain} \quad p_1' \approx p_2 \quad (19.2)$$

During the training of the adversarial framework, the *Refiner* network is guided by the feedback of the second network (*Critic*), which measures the similarity between the distribution formed by the refined samples p_1' and the distribution of the second experiment p_2. After successful training, the *Critic* network cannot find differences between the probability density p_1' and the density p_2.

The trained *Refiner* network in Fig. 19.1(a) is a mathematical representation of the quantitative differences between the probability densities of the two experiments. Physicists can explore this knowledge by analyzing the differences between the input and modified data. These differences can help, for example, to improve simulations of experiments (see Example 19.1).

Training using refined simulations: Another application is domain adaptation before training networks. In physics, most machine learning methods are trained with simulation data before applying them to measured data. This practice has been criticized for decades because the differences between measured data and simulations can lead to biases in their results that are difficult to quantify.

Refining simulated data to match measured data closely solves this problem fundamentally since the network training is performed with simulated data from the approximated probability density $p_1' \approx p_2$ (19.2) (Fig. 19.1).

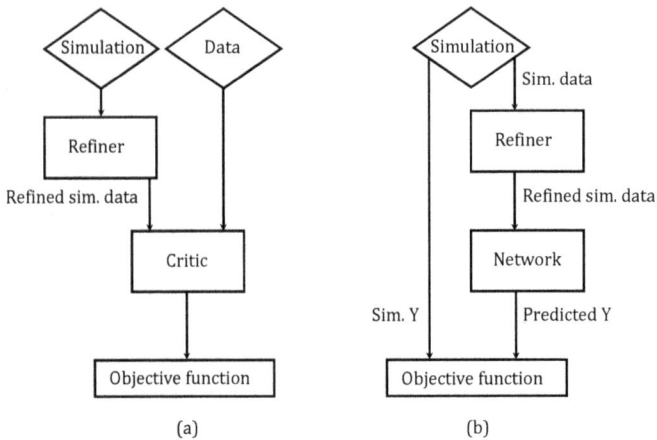

Fig. 19.1. (a) Concept of quantifying differences between data of two experiments, e.g. an experiment and its simulation. After successful training of the first network *Refiner* and the second network *Critic*, the *Critic* is unable to discriminate between simulated and measured data. (b) Training of a network to predict the physics quantity Y using data-like simulations.

Example 19.1. Signal traces of two simulation experiments: Figure 19.2 shows two different simulations together with a refined trace from one simulation, which was adapted to the other simulation.

The dotted black curve represents the signal trace as a function of time from a first simulation experiment ('simulation'). With the black full curve, a typical signal trace of a second simulation experiment is shown (referred to as 'data').

The differences between the two simulations were captured during the adversarial training [Fig. 19.1(a)] in the refiner network. These differences can be inspected by the *Refiners* modifications applied to the original simulated curves [Fig. 19.1(b)]. In Fig. 19.2 the example of an original curve ('simulation') and its modified curve ('refined simulation') is shown and compared to the second simulation ('data').

In many experiments, simulations are available which represent the relevant fundamental physics processes and describe the measured data reasonably well except for small unwanted effects. Therefore, as architecture for the *Refiner* network, residual modules are often used (see Sec. 7.5). They

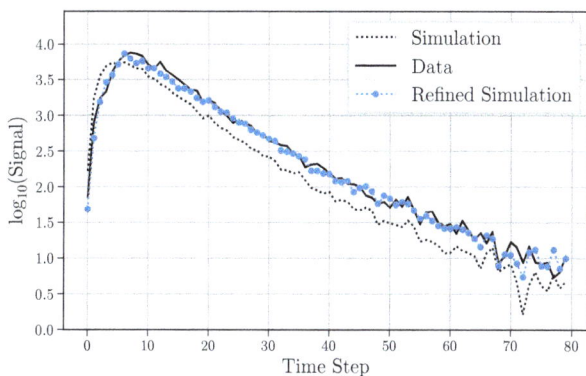

Fig. 19.2. Refinement of simulations: a signal trace of simulation 1 (dotted curve, 'simulation') has been refined (blue symbols, 'refined simulation') to follow traces of simulation 2 (black full curve). From [227] (*©Springer International Publishing AG, reprinted by permission of Springer Nature*).

allow stable training because they set a prior on the simulation and cause only small modifications. One also has to make sure that the label distributions of p_1 and p_2 match prior the training. Otherwise, during refinement, the data-label matching can be changed too much.

Example 19.2. Network training with refined signal traces: From the measurement data of a cosmic-ray experiment (see Example 11.2), the cosmic particles' energy is to be determined using a network. For this purpose, signal traces are used, as shown in Fig. 19.2.

First, the network is trained with signal traces from simulation 1. An example trace is shown as dotted black curve ('simulation') and then evaluated on signal traces from simulation 2 (full black curve, 'data'). Figure 19.3(a) shows that the predicted energy E_{DNN} of the network is skewed compared to the true simulated energy E_{true} and has a marginal resolution.

If, instead, the network is trained with the refined signal traces of simulation 1 (blue symbols 'refined simulation' in Fig. 19.2) and then applied to the traces of simulation 2, it can be seen that the energy distortion is greatly reduced. The energy resolution is much better [Fig. 19.3(b)].

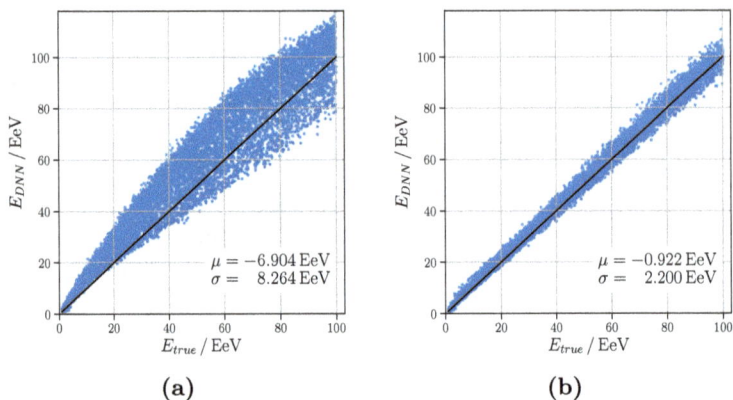

Fig. 19.3. (a) Energy reconstruction using simulation 1 for network training and evaluating on a different simulation 2. (b) Improved energy resolution after network training on the refined simulation and evaluated on simulation 2. From [227] (*©Springer International Publishing AG, reprinted by permission of Springer Nature*).

Statistical Refinement: So far, we have focused on methods that use a refiner network $f(\mathbf{x})$ to modify each data sample $\mathbf{x} \sim p_1$ individually. When applied sample-wise to the complete sample $f(p_1) \equiv p_1' \approx p_2$ is fulfilled.

Instead, we can also attempt to leave the individual data samples intact but assign them statistical weights $w(\mathbf{x})$ resulting in a weighted distribution $p_1'' = p_1(\mathbf{x})\,w(\mathbf{x})$ that approximates the target: $p_1'' \approx p_2$. The weight function should approximate the ratio of probabilities $w(\mathbf{x}) = p_2(\mathbf{x})/p_1(\mathbf{x})$.

However, both probability densities are usually available only in terms of data samples. Thus inserting a data value $\mathbf{x}$ in some functional form p_1 and p_2 is not feasible. One possible way forward is using a classification network to approximate the likelihood ratio between two samples (see Secs. 14.2 and 16.3). The classifier output can then be used as the weight function $w(\mathbf{x})$ [232].

19.2 Scale factors

One classical domain adaptation issue, present even in the absence of machine learning, arises when a labeled simulation (source domain) is used to design an algorithm or train a classifier that should be applied on experimental data (target domain) where no per-example labels are available.

Given that we can designate measurement regions in data that are relatively pure in either class, scale factors can be measured. One or two key features — features for which source and target domain show the largest variation of difference — are chosen. Then the target and source data are partitioned in these features to yield non-overlapping bins. In each bin i and for each class k the ratio of counts n is determined:

$$R_i^k = \frac{n_{i,\text{target}}^k}{n_{i,\text{source}}^k} \qquad (19.3)$$

Then, to compare distributions of arbitrary features between source and target, this ratio R can be used to reweight examples from the source distribution. For each example $\mathbf{x}$, first the bin i and class k are determined. This allows a look-up of the corresponding R_i^k value. That value is then used as multiplicative weight, e.g. when filling the example $\mathbf{x}$ in a histogram.

Scale factors are widely used, e.g. in particle physics, to correct distributions of features after selection on a classifier [233]. Of course, the determination of scale factors themselves is another problem that might be solved with machine learning approaches as it, in essence, is a regression problem [234].

While widely used and flexible, a downside of scale factors is that they only correct differences of distributions in a post-hoc way. In the following section, we will see how information from the source and target domain can be directly included as a training objective.

19.3 Common representation

If we can ensure that the latent space at an intermediate layer of a given model does not differ between source and target domains, all subsequent layers will also see no difference between the two domains. Therefore, if we can force a network only to learn shared properties relevant to a given classification task, we will have achieved domain adaptation for its final output.

A popular strategy is adversarial training,[3] illustrated in Fig. 19.4 [235]. The network architecture consists of three components: feature extractor G_f, label predictor G_y, and domain classifier G_d. The feature extractor maps input data $\mathbf{x} \in \mathbb{R}^N$ to a latent representation $f \in \mathbb{R}^M$: $f - G_f(\mathbf{x})$. This latent representation is then used as input to both label predictor and domain classifier. For training the label predictor, we employ the usual cross-entropy loss function with labeled examples from the source domain.

---

[3]We already briefly encountered adversarial training in Chapter 13 and repeat our reminder that adversarial training and adversarial examples are not related.

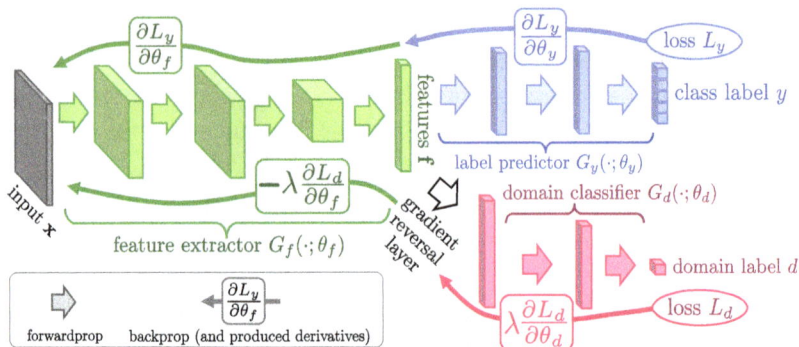

Fig. 19.4. Architecture diagram for adversarial domain adaptation [235].

Similarly, the domain classifier is trained using a cross-entropy loss to distinguish examples from the source and target domains. However, we want to reduce the difference between domains for the latent space, not increase it. Therefore, when back-propagating from the domain classifier into the feature extractor, the gradients are inverted — i.e. multiplied with a factor of minus one. Formally this is implemented in a so-called gradient reversal layer which has no trainable parameters, does not influence the forward pass of the models, but inverts the gradients in the back-propagation phase.

This technique was, for example, applied in particle physics to develop an identification algorithm for exotic long-lived particles [236]. The source domain comprises synthetic examples obtained using simulations of background processes and potential signals, while the target domain consists of experimental data. Here, adversarial domain adaptation significantly reduced the observed difference between classifier outputs on the source and target data.

19.4 Unfolding

A common problem in scientific applications is inferring some underlying truth from experimental measurements. For example, we could reconstruct a three-dimensional picture of organs and other tissue from individual images in computed tomography scans. As such tasks aim to undo a complex response function (e.g. radiation transition through materials), they are termed *inverse* problems.

We will focus on the problem of *unfolding*. Simply put, a *true* distribution is distorted — e.g. by additional physics processes or effects from limited experimental resolution — leading to an *observation* that differs from the true distribution. We then use our understanding of the distortion effects to develop a model that — on a statistical basis — infers the true distribution from the observations. In the standard formulation, we assume that for training the model, matched examples — data points for which truth and observation are available — can be used.

Mathematically, we have truth data $x_{i,\text{truth}} \in \mathbb{R}^N$ related to observed data $x_{i,\text{obs}} \in \mathbb{R}^M$ by a distorting function f:

$$f(x_{i,\text{truth}}) = x_{i,\text{obs}}. \tag{19.4}$$

In general, true and observed data can have a different number of dimensions. This already makes clear that the sought inverse function f^{-1} may not be unique. Furthermore, since often f is not given as close form expression but can even involve a chain of non-deterministic algorithms, a simple inversion is not feasible.

However, normalizing flows — introduced in Sec. 18.3 — offer a way to construct robust unfolding based on neural networks [237]. The idea is to train a bijective mapping (i.e. a flow) between the truth data and a latent space distribution of random numbers conditioned on the observed data. Importantly, this method infers the inversion f^{-1} but does not explicitly provide the forward function f as the bijective map is between random numbers and truth, not between truth and observed data. Example 19.3 explains in detail how a building block of normalizing flows can be extended for this task.

Figure 19.5 illustrates the basic idea and structure of such a conditional invertible neural network (cINN). The goal of this architecture is to unfold from observed quantities (*detector*,$\{x_d\}$) to the underlying truth (*parton*, $\{x_p\}$). Input random numbers are denoted by $\{r\}$ while a tilde indicates INN outputs. The INN g transforms bijectively between parton data and random numbers while conditioned on a learned embedding of detector inputs via the additional network f. Two loss functions are used: the latent loss $\mathcal{L}$ encoding the usual maximum Likelihood and $\mathcal{L}_{\text{MMD}}$ using maximum mean discrepancy (MMD) [238] to further improve the quality of parton-level distributions.

Alternatively, unfolding can, for example, be achieved by iteratively reweighting distributions between truth and observation until they are not

anymore distinguishable by a classifier [239]. We have already encountered — see Sec. 14.2 — that classifiers learn (a function monotonic to) the
likelihood ratio. Therefore, we can use a classifier output as a statistical
weight to modify distributions gradually. This *trick* can as well be used for
refining distributions [240].

Example 19.3. Unfolding by invertible network: In Fig. 19.6, a
variant of the previously shown invertible network block (Fig. 18.10) is
presented. Here data x_d are inserted as a *condition* to the network training and evaluation while the mapping happens between the distribution
x_p and the latent variables r.

The network training principle is the same as explained in Sec. 18.3.
We assume here that simulated data exist with the genuine distributions
x_p and corresponding observable data x_d distorted by a detector. For
training the network in the forward pass, the simulated detector data x_d
provide a condition to the network and are input as a concatenation together with the genuine values x_p. The network learns to map x_p — given
the simulated data x_d — onto normally distributed latent variables r.

In the backward pass, measured detector data x_d are inserted together
with the latent variables r and result in distributions x_p of the data
unfolded from detector effects.

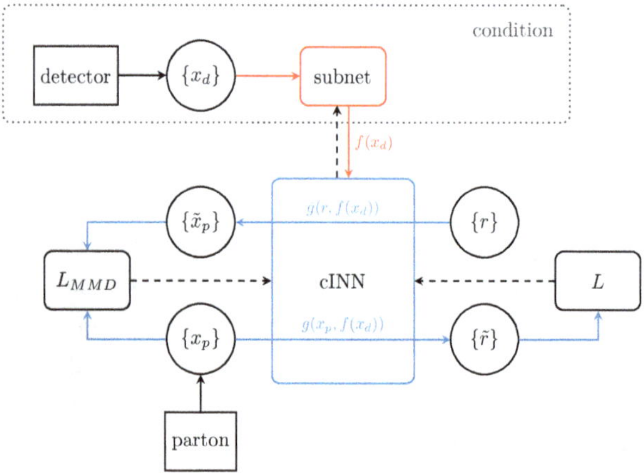

Fig. 19.5. Structure of a conditional invertible neural network (cINN)
from Ref. [237].

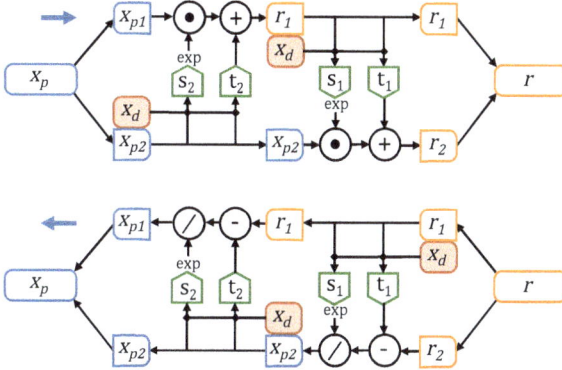

Fig. 19.6. Example for a data unfolding based on the invertible neural network shown in Fig. 18.10.

So far, we have assumed that a mapping of the true distribution and its observed counterpart is known. This requirement is not necessary, as discussed in Example 19.4. Here CycleGANs [241] are presented, initially proposed for style-transfer, to perform similarly a mapping source→ target→ source and target→ source → target and thus can be interpreted as an extended refinement concept (compare with Sec. 19.1).

Example 19.4. CycleGAN: Motivated by the frequent availability of first-principle simulations in physics, we have focused on problems with matched data — training examples for which the truth and observed view are available. However, it is also possible to construct transformations when only unmatched data — examples of both truth and observations but without one-to-one correspondence — are given. A more colorful application is style transfer of pictures, i.e. changing a photograph to look like a Monet painting or making a horse look like a zebra.

First, two adversarial frameworks similar to GANs introduced in Chapter 18, are combined to achieve this. To stay with the example of animal images: One framework is trained to transform horses into zebras (with an adversary that tries to distinguish real zebras from fake zebras), and the second framework is trained to transform zebras into horses (with an adversary that tries to distinguish real horses from fake horses).

The critical insight is then to require *cycle consistency* — adding the condition to the loss function that a chain of horse-zebra-horse should yield the original horse and zebra-horse-zebra should yield the original zebra. Such so-called *CycleGANs* can achieve several impressive image manipulations. Enforcing cycle consistency, in general, is a useful ingredient in stabilizing chains of transformations and can, therefore, be also used for refinement purposes (see Sec. 19.1).

Summary 19.1.

(1) One challenge in applying deep learning methods is substantial differences between training data and the data to be analyzed. Formally, the differences lie in the distributions of these data and the occupancy of their phase spaces by the sub-datasets of training data and application data. The various methods for solving the problem are referred to as *domain adaptation*.

(2) Using *refinement*, training data are modified in all individual event variables $x_1, \ldots, x_n$ in such a way that their formed distribution is similar to the distribution of application data. The similarity of the modified (refined) training data and application data can be proved not event-wise but in the statistical sense of ensembles. Similar label distributions are required for a successful refinement process. Besides, the method can be used to refine the training data and, thereupon, use the refined data to train networks with data that are similar to the application data.

(3) A simplified method is a *statistical* refinement, in which each training data point $\mathbf{x}_i$ is given a weight w_i so that distributions of training data and application data become indistinguishable.

(4) In the *scale factors* method, few variables x_k are selected from the data points with $x_1, \ldots, x_n$. Statistical weights are constructed from comparing their differential distributions between training and application data so that the distributions in x_k become indistinguishable. The application of this method typically happens *a posteriori*, e.g. after a classification.

(5) In the method of *the common representation*, only properties in the training data and application data for the task to be solved are used that are in agreement. All other non-matching properties are ignored.

(6) The goal of *unfolding* is to remove systematic changes to a true, natural probability distribution caused by an experimental setup. The method is an extreme form of domain adaptation, in which one returns to the original distributions. Unfolding methods are available for training data including one-to-one relations between true and observed features as well as for training data without such a one-to-one correspondence.

Chapter 20

Model independent detection of outliers and anomalies

Scope 20.1. In this chapter, we learn how to construct methods for unsupervised anomaly detection. Unsupervised anomaly detectors are models that can identify outliers, such as unexpected signals of new phenomena or unprecedented experimental malfunctions, without having access to such examples during training. We will distinguish between examples that are anomalous by themselves (point anomalies) and examples that are only notable due to their collective distributions (group anomalies). Two key approaches to build anomaly detectors will be introduced: looking for data points in regions with low background density and building likelihood ratios.

Exploring the unknown and discovering new signals is a central goal of physics, and applications range from model-independent discoveries of new phenomena to the flexible identification of unexpected hardware malfunctions. Anomaly detection is a vibrant and rapidly developing new field. While we will discuss a few fundamental principles in the following, a broader overview can, for example with a focus on particle physics, be found in Ref. [242].

With the supervised learning algorithms discussed so far, we can find 'known unknowns' — such as specific observations predicted by a particular theory or a detector problem that has been seen before. In this chapter, we turn our attention to finding 'unknown unknowns'. We will see that a clear definition becomes more challenging than for supervised problems, but it is possible to build robust anomaly detectors.

We can broadly group anomalies in two types: *individual* and *collective* outliers. Individual outliers are examples that are anomalous by themselves, for example, an invalid sensor read-out that is never encountered during regular operation or an image of a dog mixed into a sample of cat pictures. Collective outliers or group anomalies seem innocent by themselves and are only anomalies when considered in aggregate. Examples are observations belonging to a new physical process clustering in some region of phase space or a malfunction that leads to too much activity in a given sensor.

Depending on which type of anomaly is most relevant, different types of anomaly detectors exist. Specifically, we distinguish anomaly detection in regions with low background density (Sec. 20.2) from likelihood-ratio-based methods (Sec. 20.3).

20.1   Basics

We consider individual observations (e.g. images or sensor read-outs) $\mathbf{x} \in \mathbb{R}^N$ following a probability density $\mathbf{x} \sim p$. Additionally, it is assumed that at least most of these observations are generated according to a normal[1] or background process. In addition, there might exist a rare additional process called anomaly or signal. Unsupervised anomaly detection aims to detect such a signal and identify individual observations, usually without prior anomalous examples available for training.

In the following, we discuss methods to identify point anomalies and group anomalies. Point anomalies occur if the probability distributions of normal and anomalous observations are (partially) disjoint:

There exists an observation $\mathbf{x} : p(\mathbf{x}|\text{normal}) = 0$ and $p(\mathbf{x}|\text{anomaly}) > 0$

This lack of overlap leads to anomalous examples that the background process could never have produced. Figure 20.1(left) shows an example of such a point anomaly.

On the other hand, if the probability distributions overlap fully

$$\text{For all observations } \mathbf{x} : \frac{p(\mathbf{x}|\text{anomaly})}{p(\mathbf{x}|\text{normal})} < \infty$$

it is impossible to identify individual outliers. However, group anomalies — differences in the normal and anomalous densities — can still be detected. An example of a group anomaly is illustrated in Fig. 20.1(right). Of course, point anomalies might also cluster and exist as groups, allowing the combined use of approaches for both types.

---

[1]In the context of this chapter, *normal* means non-anomalous and does not refer to Gaussian distributions.

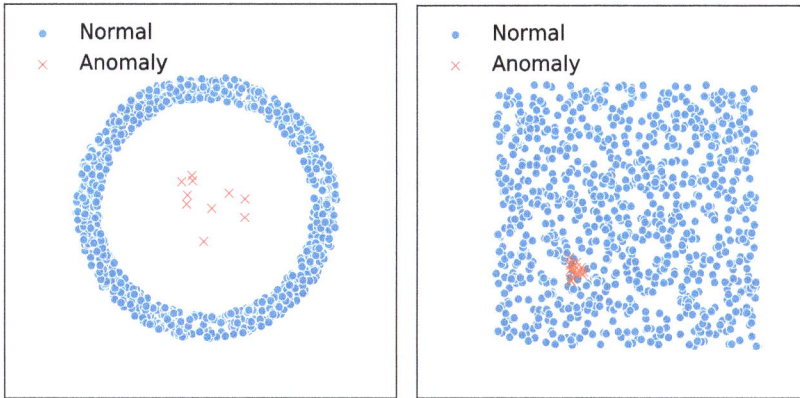

Fig. 20.1. Example of several point anomalies (left) and a group anomaly (right).

Assuming high-quality simulations of both the normal data and potential anomalies are available, it turns this problem into a simple supervised classification task. In the context of anomaly detection, we assume that no simulated signal is available. Lack of signal examples makes potential solutions more sensitive to a broader range of anomalies but more challenging to develop.

However, for constructing anomaly detection methods, we will sometimes require a data set of normal events that we can consider to be free of anomalies. Such data could either be obtained using a simulation of the normal process or by performing a control measurement. We label this additional data $\tilde{p}$ and assume

$$\tilde{p}(\mathbf{x}) = \tilde{p}(\mathbf{x}|\text{normal}) = p(\mathbf{x}|\text{normal}). \tag{20.1}$$

The first equality states that the control measurement does not contain anomalies, the second one that normal events from the control measurement behave like the data we test for the presence of anomalies.

In practice, both assumptions might hold only approximately. Violation of the first one is termed *pollution* and occurs when control measurement cannot be guaranteed to be free of anomalies. Violation of the second equality is a typical problem of simulation based methods and an example where domain adaptation (see Chapter 19) might be used.

We will limit our discussion in the following to building anomaly detectors — i.e. functions that assign a score to each observation $f(\mathbf{x})$ so that high scores correspond to more anomalous events, either further away from the background or in regions of anomalous density. For the practical application of $f(\mathbf{x})$, additional work is needed: essentially, a statistical model that calibrates the anomaly score and yields the significance associated with a set of observations. Typical approaches, again, either rely on simulation or use so-called control regions defined in data.

20.2 Low background density

If data known to contain only normal events $\tilde{\mathbf{x}} \sim \tilde{p}$ — e.g. obtained using simulation or a control measurement — is available, anomaly detection could be implemented by comparing it to actual data classifying all differences as anomalous.

However, in realistic problems, the feature space is typically high dimensional, and densities are not available in a tractable way, but only implicitly defined in the form of examples $\tilde{\mathbf{x}} \sim \tilde{p}$ and $\mathbf{x} \sim p$, respectively. Nevertheless, a systematic comparison of simulation and observed data can still be made. Even without machine learning, it is possible to compare histogrammed features bin-by-bin [243, 244]. While straightforward, this method will have problems identifying anomalies that manifest only as correlations of several features.

A popular approach for anomaly detection that implicitly finds regions of low background density relies on autoencoders [245] (see Chapter 17). The learning target of autoencoder networks is to compress (and decompress) input data by minimizing a reconstruction loss such as $|\mathbf{x} - f(\mathbf{x})|$. If such a network is trained only on normal samples, it learns to compress and decompress them well. However, this will not work for anomalous examples never encountered during training — especially point anomalies — leading to a larger reconstruction loss. This makes the loss of a trained autoencoder an interesting anomaly score [246].

The choice of compression strength is a crucial hyperparameter in designing such an approach. If the capacity and the latent space size are sufficient to learn the identity function, no compression is needed, and no anomaly detection will be possible. On the other hand, too small latent spaces will limit the expressiveness of the network and lead to high reconstruction losses for normal and anomalous events alike. Another potential failure mode are anomalies that are inherently less complex than normal examples — in such cases, compression still might be easier and lead to a low reconstruction loss, even if such examples were not seen during training.

An approach that directly encodes the goal of anomaly detection in the learning objective is Deep Support Vector Data Description (Deep SVDD) networks [247]. Here, the objective function rewards mapping all inputs as close to a central point as possible. The trivial solution — a network that maps all inputs to the same central point c — is prevented by specific architecture choices. For example, no bias nodes are permitted. Once trained, the distance from c can again be used to detect point anomalies.

20.3 Likelihood-ratio based detection

The methods discussed so far had in common that they were trained solely on normal events to obtain an anomaly score that implicitly — as in the case of autoencoders — or explicitly — as for Deep SVDD — identified regions of low density p. However, we know that a likelihood ratio yields the most powerful test statistic (see Sec. 14.2). Unfortunately, in unsupervised anomaly detection, we do not have anomalous examples for building a signal-to-background (S/B) likelihood ratio

$$L_{S/B} = \frac{p(\mathbf{x}|\text{anomaly})}{p(\mathbf{x}|\text{normal})}. \tag{20.2}$$

Nevertheless, we can construct another useful ratio using the additional anomaly-free data $\tilde{p}$, namely the data-to-background likelihood-ratio (D/B)

$$L_{D/B} = \frac{p(\mathbf{x})}{\tilde{p}(\mathbf{x}|\text{normal})}. \tag{20.3}$$

In the following, we first motivate why this ratio is indeed helpful in identifying anomalies detection and then sketch two methods for how it can be used.

Even without knowing the numerical values of the relative fractions of normal (f_{normal}) and anomalous (f_{anomaly}) examples in the data, we can still write them down. We assume unity:

$$f_{\text{normal}} + f_{\text{anomaly}} = 1$$

Note that this does not imply that an anomaly has to be present, f_{anomaly} could simply be zero.

This trick allows us to write down the two components (normal and anomaly) of the data probability distribution separately:

$$p(\mathbf{x}) = f_{\text{normal}}\, p(\mathbf{x}|\text{normal}) + f_{\text{anomaly}}\, p(\mathbf{x}|\text{anomaly}) \tag{20.4}$$

Inserting this into (20.3) and using (20.1) shows that $L_{D/B}$ is indeed a good anomaly score

$$L_{D/B} = f_{\text{normal}} + f_{\text{anomaly}} \frac{p(\mathbf{x}|\text{anomaly})}{p(\mathbf{x}|\text{normal})}, \tag{20.5}$$

as it is monotonous to $L_{S/B}$ (20.2). If two functions are monotonous to each other, they have the same performance as a classifier with the only difference that the numerical threshold values change.

In the following, we will discuss two methods that use (20.3): label-noise-based and density-estimation-based anomaly detection.

We have already observed (see Sec. 16.3) that the classification of impure samples is possible. In this case, we can train a binary classifier to distinguish samples drawn from the data p against the anomaly-free data $\tilde{p}$. Following the reasoning above, a selection using the classifier output should enhance anomalies present in the data.

A concrete proposal for label-noise-based anomaly detection is the so-called CWoLa-Hunting method [248]. Here non-overlapping selections in one feature are chosen. The data passing some selection (e.g. a certain window in one feature) are tested for the presence of anomalies and take the role of p. Data failing this selection (e.g. the region outside the selected window) are considered anomaly free $\tilde{p}$ and are often termed side-band. For a classifier-based approach to work, the possible presence of an anomaly must be the only difference between the selected data and the side-band distribution. That means observables correlated with the feature used to define the side-bands cannot be used as training input, which yields a limitation of this approach. In practice, one of course, can move the window and define multiple selections with corresponding side-bands and perform a test for anomalies in all of them.

While CWoLa-Hunting is straightforward to implement, its performance is constrained by the limited choice of input features to avoid correlations. To circumvent this constraint, two conditional density estimators q (see Sec. 18.3) can be trained instead, one to replicate $p \approx q_{\text{inner}}$ and one on $\tilde{p} \approx q_{\text{outer}}$. By extrapolating the density estimation from the anomaly-free data to the data where we look for anomalies, the ratio

$$R = \frac{q_{\text{inner}}}{q_{\text{outer}}} \tag{20.6}$$

again results in a function monotonous to the likelihood ratio. It therefore yields a promising and powerful density-estimation-based approach for anomaly detection [214].

20.4 Exercises

Exercise 20.1. We will use a simple, autoencoder-based, anomaly detector on pictures of clothing items from the Fashion-MNIST dataset. This dataset contains 28×28 pixel grayscale images of ten different types of clothing. We will use training data from one of these classes and see if we can identify the others as anomalies.

(1) Use the notebook (Exercise 20.1 from the exercise page) to access the data and additional material.
(2) Either use the provided implementation or develop your own autoencoder architecture to compress/decompress your favorite clothing item.
(3) Write a function that calculates the squared, pixel-wise difference between input images and the output of the autoencoder for each image.
(4) Plot the normalized distribution of this observable for the clothing item used for training as well as for the other classes. Can this be used as anomaly detector?

Summary 20.1.

(1) Anomaly detection deals with the problem of identifying potential outliers without having access to specific examples during the training phase.

(2) We distinguish between point anomalies (individual outliers) and group anomalies (collective outliers).

(3) Two principal methods for anomaly detection can be used:

- Identifying examples in regions with a low density of normal examples, e.g. by using autoencoders trained purely on background examples.
- Constructing a data-to-background likelihood ratio (for example, by estimating the background from control measurements) that acts as a proxy for a signal-to-background likelihood ratio.

Chapter 21

Beyond the scope of this textbook

Developments in the field of artificial intelligence and especially machine learning proceed at a rapid pace. Therefore, this chapter carries a particular risk of being outdated already tomorrow. In our own research, machine learning methods are a means to an end of gaining deeper insights into our physics research topics. Our following compilation of related areas and recent developments is therefore incomplete and with a personal weighting.

First, we need to point out that this book focused on ideas and concepts. Accordingly, we sidestepped practical topics such as: building data processing pipelines; working with very large or distributed datasets; efficiently training and evaluating machine learning models at scale; or finding optimal trade-offs between speed and memory consumption. These issues are, of course, crucially important, and we refer to Ref. [249] for an overview of currently available technologies.

In the past decade, a dramatic increase in power consumption for computing has been observed worldwide. Simultaneously, there is a strong commitment to perform data-driven computations on state-of-the-art devices and even on the smallest devices. Intensive research is being conducted on energy-efficient new hardware including quantum computers and memory- and energy-efficient algorithms in these contexts.

21.1 Dedicated hardware and microsystems

One strategy is to use dedicated hardware. In the transition from the CPU (Central Processing Unit) to the GPU (Graphics Processing Unit) for parallel-running calculations in network training, the energy advantage due to the time factor is evident with similar power consumption. Modern advancements such as the TPU (Tensor Processing Unit) contain hardware units, e.g. for matrix multiplication, and thus achieve additional acceleration.

Intensive research is also being performed on hardware reduction to microsystems such as FPGA (Field Programmable Gate Array) or ASIC (Application-Specific Integrated Circuit). These developments are also being driven by detector physicists, who require ultra-fast decisions close to measurement sensors while using minimal energy. To bring deep neural networks to work here, a massive reduction in memory requirements is needed. To achieve this, the number of bits used for the weight parameters is reduced, in the extreme case of binary networks, to one bit only [250, 251]. Only slightly more relaxed are ternary neural networks, which allow three values per weight parameter [252]. For user programming of neural networks on FPGAs, there is convenient software for converting the networks into the FPGAs' assembly language [253].

Furthermore, by using qubits in quantum computing instead of bits, extreme acceleration of computations is expected. This advantage would also be apparent in quantum machine learning [254, 255].

21.2 Neuroscience inspired developments

One argument is provided by neuroscience. The human brain consumes only a tiny fraction of the energy for recognizing objects compared to a computer. Therefore, one branch of research focuses on the brain's structural functionalities and their mapping onto novel hardware [256–258].

Two aspects are particularly relevant for improving energy efficiency. In von Neumann's computer architecture, memory and computing units are separate components. To save the two-way transport of information, research is being carried out on systems that merge memory and computing unit. One example is memristors in matrix structures (computation), whose resistances can be adjusted depending on the voltage (memory). These could greatly accelerate computations with neural networks [259].

The second energy-saving aspect is the information transport itself, which is realized similarly to the brain by bundles of temporally successive short pulses. So-called Spiking Neural Networks have been constructed based on short pulses. Both forward computation and backpropagation for training are challenges here [260, 261]. In the meantime, benchmark tests of Spiking Neural Networks have been performed for all common standard architectures (see Part 2). Their results are similar to those of standard neural networks [262].

21.3 Information field theory

A completely different idea beyond this book is machine learning techniques that do not require a large amount of training data. An extreme example is a single astronomical image with a large number of stars to be enumerated while being obscured by diffuse light. In the so-called information field theory, instead of much training data, the parameters of the numerical method to be trained are bound using physical assumptions [263]. Posterior distributions of the target variable are calculated using Bayes' theorem, with assumptions formulated in prior functions. For example, experience shows that delta functions rarely arise in nature. Such an assumption of smoothness can be formulated in terms of 2-point correlation functions.

21.4 Human readable physics concepts

Finally, we highlight another aspect. As physicists, we are trained in the mathematical description of nature phenomena. Now we train neural networks to record properties and correlations between properties of data in great detail. How do we retrieve this stored knowledge from the network? For example, experiments are being conducted with the specification of analytic functions whose weights are autonomously assigned by a neural network based on the data [264]. Autonomous recognition of physics concepts from measurement data in human-readable mathematical formulations would have far-reaching implications for physics research in general.

Appendix A

Notations

Table A.1. Notations used in this textbook.

Notation	
(1.1)	reference to equation number
$\vec{x}$	vector
$\mathbf{x}$	data (vector, matrix, tensor,...)
$\mathbf{A}$	matrix
$\mathbf{W}, W_{ij}$	weight matrix, single weight
θ, φ	adaptive parameter
$\mathcal{L}$	loss, cost, objective function
g, f	functions
p	probability, probability density
$\mathbf{x} \sim p(\mathbf{x})$	random variable has distribution p
$\sigma(x)$	activation function
$\mathbb{E}$	expectation value
$\vec{x} \cdot \vec{y}$	scalar product of $\vec{x}$ and $\vec{y}$
$\mathbf{A} \odot \mathbf{B}$	element-wise (Hadamard) product of $\mathbf{A}$ and $\mathbf{B}$

Bibliography

[1] I. Goodfellow, Y. Bengio and A. Courville, *Deep Learning*. (MIT Press, 2016), http://www.deeplearningbook.org.

[2] M. Feickert and B. Nachman, A living review of machine learning for particle physics, (2021), arXiv:2102.02770 [hep-ph].

[3] A. Tanaka, A. Tomiya and K. Hashimoto, *Deep Learning and Physics*. (Springer International Publishing, 2021), doi:10.1007/978-981-33-6108-9, https://www.springer.com/gp/book/9789813361072.

[4] J. R. Koza, F. H. Bennett, D. Andre and M. A. Keane, *Automated Design of Both the Topology and Sizing of Analog Electrical Circuits Using Genetic Programming*, pp. 151–170. (Springer Netherlands, Dordrecht, 1996), doi: 10.1007/978-94-009-0279-4_9.

[5] Y. LeCun, Deep learning, (2014), https://cilvr.nyu.edu/lib/exe/fetch.php?media=deeplearning:dl-intro.pdf.

[6] G. Van Rossum and F. L. Drake, *Python 3 Reference Manual*. CreateSpace, Scotts Valley, CA (2009), ISBN 1441412697.

[7] F. Chollet *et al.*, Keras, (2015), https://github.com/fchollet/keras.

[8] F. Chollet, *Deep Learning with Python*. (Manning Publications Co., 2017).

[9] A. Gulli and S. Pal, *Deep learning with Keras*. (Packt Publishing Ltd., 2017).

[10] M. Abadi *et al.*, Tensorflow: Large-scale machine learning on heterogeneous distributed systems, (2016), arXiv:1603.04467 [cs.DC].

[11] A. Paszke *et al.*, *PyTorch: An Imperative Style, High-Performance Deep Learning Library*. (Curran Associates, Inc., 2019), arXiv:1912.01703 [cs.LG].

[12] K. Hornik, M. Stinchcombe and H. White, Multilayer feedforward networks are universal approximators, *Neural Networks* **2**, 5, pp. 359–366 (1989), doi:https://doi.org/10.1016/0893-6080(89)90020-8.

[13] G. Cybenko, Approximation by superpositions of a sigmoidal function, *Mathematics of Control, Signals and Systems* **2**, pp. 303–314 (1989).

[14] K. Hornik, Approximation capabilities of multilayer feedforward networks, *Neural Networks* **4**, 2, pp. 251–257 (1991), doi:https://doi.org/10.1016/0893-6080(91)90009-T.

[15] B. Hanin and M. Sellke, Approximating continuous functions by ReLU Nets of minimal width, (2017), `arXiv:1710.11278 [stat.ML]`.

[16] Z. Lu, H. Pu, F. Wang, Z. Hu and L. Wang, The expressive power of neural networks: A view from the width, in I. Guyon *et al.* (eds.), *Advances in Neural Information Processing Systems 30*. Curran Associates, Inc., pp. 6231–6239 (2017).

[17] X. Glorot and Y. Bengio, Understanding the difficulty of training deep feedforward neural networks, *Journal of Machine Learning Research — Proceedings Track* **9**, pp. 249–256 (2010).

[18] K. He, X. Zhang, S. Ren and J. Sun, Delving deep into rectifiers: Surpassing human-level performance on imagenet classification, *CoRR* (2015), `arXiv:1502.01852`.

[19] A. LeNail, Nn-svg: Publication-ready neural network architecture schematics, *Journal Open Source Software* **4**, p. 747 (2019).

[20] J. Duchi, E. Hazan and Y. Singer, Adaptive subgradient methods for online learning and stochastic optimization, *Journal of Machine Learning Research* **12**, 61, pp. 2121–2159 (2011), `http://jmlr.org/papers/v12/duchi11a.html`.

[21] G. Hinton, Lecture series, lecture 6e in 2014, (2012), `http://www.cs.toronto.edu/~tijmen/csc321`, accessed 22-Jan-2021.

[22] B. Polyak, Some methods of speeding up the convergence of iteration methods, *USSR Computational Mathematics and Mathematical Physics* **4**, 5, pp. 1–17 (1964), doi:https://doi.org/10.1016/0041-5553(64)90137-5.

[23] D. P. Kingma and J. Ba, Adam: A method for stochastic optimization, *arXiv e-prints* (2014), `arXiv:1412.6980 [cs.LG]`.

[24] S. J. Reddi, S. Kale and S. Kumar, On the convergence of Adam and beyond, (2019), `arXiv:1904.09237 [cs.LG]`.

[25] C. Sammut and G. I. Webb (eds.), *Encyclopedia of Machine Learning and Data Mining*, 2nd edn., Springer Reference. (Springer, New York, 2017), ISBN 978-1-4899-7685-7, doi:10.1007/978-1-4899-7687-1.

[26] B. Xu, N. Wang, T. Chen and M. Li, Empirical evaluation of rectified activations in convolutional network, (2015), `arXiv:1505.00853`.

[27] N. Langner, Private Communication (2021), Phys. Inst. 3A, RWTH Aachen University, 52056 Aachen, Germany.

[28] S. Ioffe and C. Szegedy, Batch normalization: Accelerating deep network training by reducing internal covariate shift, *Proceedings of the 32nd International Conference on Machine Learning* **37**, pp. 448–456 (2015), `arXiv:1502.03167`.

[29] K. He, X. Zhang, S. Ren and J. Sun, Deep residual learning for image recognition, (2015), `arXiv:1512.03385 [cs.CV]`.

[30] C. M. Bishop, *Neural Networks for Pattern Recognition*. (Oxford University Press, Inc., USA, 1995), ISBN 0198538642.

[31] B. D. Ripley, *Pattern Recognition and Neural Networks*. (Cambridge University Press, 1996), doi:10.1017/CBO9780511812651.

[32] W. N. Venables and B. D. Ripley, *Modern Applied Statistics with S*. (Springer, New York, 2002), ISBN 0-387-95457-0.

[33] R. K. Srivastava, K. Greff and J. Schmidhuber, Highway networks, (2015), arXiv:1505.00387 [cs.LG].

[34] D.-A. Clevert, T. Unterthiner and S. Hochreiter, Fast and accurate deep network learning by exponential linear units (elus), *CoRR* (2016), arXiv:1511.07289 [cs.LG].

[35] G. Klambauer, T. Unterthiner, A. Mayr and S. Hochreiter, Self-normalizing neural networks, *arXiv e-prints* (2017), arXiv:1706.02515 [cs.LG].

[36] Y. Lecun, L. Bottou, Y. Bengio and P. Haffner, Gradient-based learning applied to document recognition, *Proceedings of the IEEE* **86**, 11, pp. 2278–2324 (1998).

[37] A. Krizhevsky, I. Sutskever and G. E. Hinton, Imagenet classification with deep convolutional neural networks, in F. Pereira *et al.* (eds.), *Advances in Neural Information Processing Systems 25*. Curran Associates, Inc., pp. 1097–1105 (2012).

[38] M. Morgenstern, Private Communication (2021), Phys. Inst. 2B, RWTH Aachen University, 52056 Aachen, Germany.

[39] R. Wiesendanger, *Scanning Probe Microscopy and Spectroscopy*. (Cambridge University Press, Cambridge, 1994).

[40] C. Szegedy, W. Liu, Y. Jia, P. Sermanet, S. Reed, D. Anguelov, D. Erhan, V. Vanhoucke and A. Rabinovich, Going deeper with convolutions, *2015 IEEE Conference on Computer Vision and Pattern Recognition (CVPR)*, pp. 1–9 (2015).

[41] J. Hu, L. Shen, S. Albanie, G. Sun and E. Wu, Squeeze-and-excitation networks, (2019), arXiv:1709.01507 [cs.CV].

[42] A. G. Howard *et al.*, Mobilenets: Efficient convolutional neural networks for mobile vision applications, (2017), arXiv:1704.04861 [cs.CV].

[43] K. He, X. Zhang, S. Ren and J. Sun, Deep residual learning for image recognition, in *2016 IEEE Conference on Computer Vision and Pattern Recognition (CVPR)*, pp. 770–778 (2016).

[44] G. Huang, Z. Liu, L. Van Der Maaten and K. Q. Weinberger, Densely connected convolutional networks, in *2017 IEEE Conference on Computer Vision and Pattern Recognition (CVPR)*, pp. 2261–2269 (2017).

[45] O. Ronneberger, P. Fischer and T. Brox, U-net: Convolutional networks for biomedical image segmentation, (2015), arXiv:1505.04597 [cs.CV].

[46] E. Hoogeboom, J. W. T. Peters, T. S. Cohen and M. Welling, Hexaconv, *CoRR* (2018), arXiv:1803.02108, http://arxiv.org/abs/1803.02108.

[47] A. Aurisano *et al.*, A convolutional neural network neutrino event classifier, *Journal of Instrumentation* **11**, 09 (2016), doi:10.1088/1748-0221/11/09/p09001, http://dx.doi.org/10.1088/1748-0221/11/09/P09001.

[48] P. T. Komiske, E. M. Metodiev and M. D. Schwartz, Deep learning in color: towards automated quark/gluon jet discrimination, *Journal of High Energy Physics* **2017**, 1 (2017), doi:10.1007/jhep01(2017)110, http://dx.doi.org/10.1007/JHEP01(2017)110.

[49] C. Adams *et al.*, Deep neural network for pixel-level electromagnetic particle identification in the microboone liquid argon time projection chamber, *Physical Review D* **99**, 9 (2019), doi:10.1103/physrevd.99.092001, http://dx.doi.org/10.1103/PhysRevD.99.092001.

[50] L. Schermelleh, A. Ferrand, T. Huser, C. Eggeling, M. Sauer, O. Biehlmaier and G. P. C. Drummen, Super-resolution microscopy demystified, *Nature Cell Biology* **21**, pp. 72–84 (2019).

[51] T. Aritake, H. Hino, S. Namiki, D. Asanuma, K. Hirose and N. Murata, Single-molecule localization by voxel-wise regression using convolutional neural network, *Results in Optics* **1**, pp. 100019–100029 (2020).

[52] D. Rumelhart, G. E. Hinton and R. Williams, Learning representations by back-propagating errors, *Nature* **323**, pp. 533–536 (1986).

[53] C. Olah, Understanding LSTM networks, (2015), `colah.github.io/posts/2015-08-Understanding-LSTMs/`.

[54] H. Siegelmann and E. Sontag, On the computational power of neural nets, *Journal Computer and Systems Sciences* **50**, pp. 132–150 (1995).

[55] S. Hochreiter and J. Schmidhuber, Long short-term memory, *Neural Computation* **9**, pp. 1735–1785 (1997).

[56] R. Józefowicz, W. Zaremba and I. Sutskever, An empirical exploration of recurrent network architectures, in *ICML* (2015).

[57] F. Gers, J. Schmidhuber and F. Cummins, Learning to forget: continual prediction with LSTM, in *9th International Conference on Artificial Neural Networks: ICANN '99*, pp. 850–855 (1999).

[58] K. Greff, R. K. Srivastava, J. Koutník, B. R. Steunebrink and J. Schmidhuber, LSTM: A search space odyssey, *IEEE Transactions on Neural Networks and Learning Systems* **28**, pp. 2222–2232 (2015).

[59] F. Gers, N. Schraudolph and J. Schmidhuber, Learning precise timing with LSTM recurrent networks, *Journal of Machine Learning Research* **3**, pp. 115–143 (2002).

[60] K. Cho, B. van Merrienboer, C. Gulcehre, D. Bahdanau, F. Bougares, H. Schwenk and Y. Bengio, Learning phrase representations using RNN encoder-decoder for statistical machine translation, (2014), `arXiv:1406.1078v3 [cs.CL]`.

[61] J. Chung, C. Gulcehre, K. Cho and Y. Bengio, Empirical evaluation of gated recurrent neural networks on sequence modeling, (2014), `arXiv:1412.3555 [cs.NE]`.

[62] G.-B. Zhou, J. Wu, C.-L. Zhang and Z.-H. Zhou, Minimal gated unit for recurrent neural networks, (2016), `arXiv:1603.09420 [cs.NE]`.

[63] S. Li, J. Chen and B. Liu, Protein remote homology detection based on bidirectional long short-term memory, *BMC Bioinformatics* **18**, pp. 443–536 (2017).

[64] M. Wielgosz, A. Skoczen and M. Mertik, Using LSTM recurrent neural networks for monitoring the LHC superconducting magnets, *Nuclear Inst. and Methods in Physics Research A* **867**, pp. 40–50 (2017).

[65] A. Mehta, C. Scott, D. Oyen, N. Panda and G. Srinivasan, Physics-informed spatiotemporal deep learning for emulating coupled dynamical systems, in *Proceedings of the AAAI 2020 Spring Symposium on Combining Artificial Intelligence and Machine Learning with Physical Sciences*, Vol. 2587 (2020), `http://ceur-ws.org/Vol-2587/article_11.pdf`.

[66] The IceCube Collaboration, M. G. Aartsen *et al.*, The IceCube Neutrino

Observatory Part II: Atmospheric and Diffuse UHE Neutrino Searches of All Flavors, (2014), `arXiv:1309.7003 [astro-ph.HE]`.

[67] The CMS Collaboration, CMS Event displays displaying the use of the tau embedding technique, (2020), `https://cds.cern.ch/record/2711418`, CMS Collection.

[68] Y. Akrami *et al.*, Planck 2018 results. I. Overview and the cosmological legacy of Planck, *Astronomy and Astrophysics* **641**, p. 1 (2018).

[69] M. Gori, G. Monfardini and F. Scarselli, A new model for learning in graph domains, in *Proceedings. 2005 IEEE International Joint Conference on Neural Networks, 2005*, Vol. 2, pp. 729–734 vol. 2 (2005), doi: 10.1109/IJCNN.2005.1555942.

[70] F. Scarselli, M. Gori, A. C. Tsoi, M. Hagenbuchner and G. Monfardini, The graph neural network model, *IEEE Transactions on Neural Networks* **20**, 1, pp. 61–80 (2009), doi:10.1109/TNN.2008.2005605.

[71] D. Grattarola and C. Alippi, Graph neural networks in tensorflow and keras with spektral, (2020), `arXiv:2006.12138 [cs.LG]`.

[72] M. Fey and J. E. Lenssen, Fast graph representation learning with PyTorch Geometric, in *ICLR Workshop on Representation Learning on Graphs and Manifolds* (2019).

[73] M. Wang *et al.*, Deep graph library: A graph-centric, highly-performant package for graph neural networks, (2020), `arXiv:1909.01315 [cs.LG]`.

[74] S. G. Kobourov, Spring embedders and force directed graph drawing algorithms, (2012), `arXiv:1201.3011 [cs.CG]`.

[75] A. A. Hagberg, D. A. Schult and P. J. Swart, Exploring network structure, dynamics, and function using networkx, in G. Varoquaux, T. Vaught and J. Millman (eds.), *Proceedings of the 7th Python in Science Conference*. Pasadena, CA USA, pp. 11–15 (2008).

[76] M. M. Bronstein, J. Bruna, Y. LeCun, A. Szlam and P. Vandergheynst, Geometric deep learning: Going beyond euclidean data, *IEEE Signal Processing Magazine* **34**, 4, pp. 18–42 (2017), doi:10.1109/MSP.2017.2693418.

[77] T. N. Kipf and M. Welling, Semi-supervised classification with graph convolutional networks, (2017), `arXiv:1609.02907 [cs.LG]`.

[78] F. Monti, D. Boscaini, J. Masci, E. Rodolà, J. Svoboda and M. M. Bronstein, Geometric deep learning on graphs and manifolds using mixture model cnns, (2016), `arXiv:1611.08402 [cs.CV]`.

[79] K. Xu, W. Hu, J. Leskovec and S. Jegelka, How powerful are graph neural networks? (2019), `arXiv:1810.00826 [cs.LG]`.

[80] W. L. Hamilton, R. Ying and J. Leskovec, Representation learning on graphs: Methods and applications, *CoRR* (2017), `arXiv:1709.05584`.

[81] J. Shlomi, P. Battaglia and J.-R. Vlimant, Graph neural networks in particle physics, *Machine Learning: Science and Technology* **2**, 2, p. 021001 (2021), doi:10.1088/2632-2153/abbf9a, `http://dx.doi.org/10.1088/2632-2153/abbf9a`.

[82] J. Gilmer, S. S. Schoenholz, P. F. Riley, O. Vinyals and G. E. Dahl, Neural message passing for quantum chemistry, (2017), `arXiv:1704.01212 [cs.LG]`.

[83] W. W. Zachary, An information flow model for conflict and fission in small groups, *Journal of Anthropological Research* **33**, 4, pp. 452–473 (1977), http://www.jstor.org/stable/3629752.

[84] D. Duvenaud *et al.*, Convolutional networks on graphs for learning molecular fingerprints, (2015), arXiv:1509.09292 [cs.LG].

[85] M. Simonovsky and N. Komodakis, Dynamic edge-conditioned filters in convolutional neural networks on graphs, (2017), arXiv:1704.02901 [cs.CV].

[86] H. Qu and L. Gouskos, Jet tagging via particle clouds, *Physical Review D* **101**, 5 (2020), doi:10.1103/physrevd.101.056019, http://dx.doi.org/10.1103/PhysRevD.101.056019.

[87] Y. Wang, Y. Sun, Z. Liu, S. E. Sarma, M. M. Bronstein and J. M. Solomon, Dynamic graph CNN for learning on point clouds, (2019), arXiv:1801.07829 [cs.CV].

[88] S. R. Qasim, J. Kieseler, Y. Iiyama and M. Pierini, Learning representations of irregular particle-detector geometry with distance-weighted graph networks, *The European Physical Journal C* **79**, 7 (2019), doi:10.1140/epjc/s10052-019-7113-9, http://dx.doi.org/10.1140/epjc/s10052-019-7113-9.

[89] T. Bister, M. Erdmann, J. Glombitza, N. Langner, J. Schulte and M. Wirtz, Identification of patterns in cosmic-ray arrival directions using dynamic graph convolutional neural networks, *Astroparticle Physics* **126**, p. 102527 (2021), doi:10.1016/j.astropartphys.2020.102527, http://dx.doi.org/10.1016/j.astropartphys.2020.102527.

[90] G. Kasieczka *et al.*, The machine learning landscape of top taggers, *SciPost Phys.* **7**, p. 14 (2019), doi:10.21468/SciPostPhys.7.1.014, https://scipost.org/10.21468/SciPostPhys.7.1.014.

[91] M. Zaheer, S. Kottur, S. Ravanbakhsh, B. Poczos, R. Salakhutdinov and A. Smola, Deep sets, *NeurIPS 2017* (2017), arXiv:1703.06114.

[92] P. T. Komiske, E. M. Metodiev and J. Thaler, Energy flow networks: deep sets for particle jets, *Journal of High Energy Physics* **2019**, 1 (2019), doi:10.1007/jhep01(2019)121, http://dx.doi.org/10.1007/JHEP01(2019)121.

[93] J. Bruna, W. Zaremba, A. Szlam and Y. LeCun, Spectral networks and locally connected networks on graphs, (2014), arXiv:1312.6203 [cs.LG].

[94] G. Arfken, Mathematical methods for physicists, *American Journal of Physics* **35**, 11, pp. 1097–1098 (1967), doi:10.1119/1.1973757, https://doi.org/10.1119/1.1973757.

[95] D. I. Shuman, B. Ricaud and P. Vandergheynst, Vertex-frequency analysis on graphs, (2013), arXiv:1307.5708 [math.FA].

[96] M. Saerens, F. Fouss, L. Yen and P. Dupont, The principal components analysis of a graph, and its relationships to spectral clustering, in J.-F. Boulicaut *et al.* (eds.), *Machine Learning: ECML 2004*, pp. 371–383. (Springer, Berlin, Heidelberg, 2004), ISBN 978-3-540-30115-8.

[97] M. Henaff, J. Bruna and Y. LeCun, Deep convolutional networks on graph-structured data, (2015), arXiv:1506.05163 [cs.LG].

[98] M. Defferrard, X. Bresson and P. Vandergheynst, Convolutional neural networks on graphs with fast localized spectral filtering, (2017), `arXiv:1606.09375 [cs.LG]`.

[99] D. K. Hammond, P. Vandergheynst and R. Gribonval, Wavelets on graphs via spectral graph theory, (2009), `arXiv:0912.3848 [math.FA]`.

[100] K. Górski *et al.*, Healpix: A framework for high-resolution discretization and fast analysis of data distributed on the sphere, *The Astrophysical Journal* **622**, pp. 759–771 (2005).

[101] N. Perraudin, M. Defferrard, T. Kacprzak and R. Sgier, Deepsphere: Efficient spherical convolutional neural network with healpix sampling for cosmological applications, *Astronomy and Computing* **27**, p. 130–146 (2019), doi:10.1016/j.ascom.2019.03.004, `http://dx.doi.org/10.1016/j.ascom.2019.03.004`.

[102] T. S. Cohen, M. Geiger, J. Koehler and M. Welling, Spherical cnns, (2018), `arXiv:1801.10130 [cs.LG]`.

[103] S. Ruder, An overview of multi-task learning in deep neural networks, *CoRR* (2017), `arXiv:1706.05098`.

[104] A. Aab *et al.*, Deep-Learning based reconstruction of the shower maximum X_{max} using the water-Cherenkov detectors of the Pierre Auger Observatory, (2021), `arXiv:2101.02946 [astro-ph.IM]`.

[105] W. Samek and K. Müller, Towards explainable artificial intelligence, *Lecture Notes in Computer Science*, pp. 5–22 (2019), doi:10.1007/978-3-030-28954-6_1, `http://dx.doi.org/10.1007/978-3-030-28954-6_1`.

[106] N. Xie, G. Ras, M. van Gerven and D. Doran, Explainable deep learning: A field guide for the uninitiated, (2020), `arXiv:2004.14545 [cs.LG]`.

[107] M. D. Zeiler and R. Fergus, Visualizing and understanding convolutional networks, (2013), `arXiv:1311.2901 [cs.CV]`.

[108] Tensorflow/lucid, (2021), `https://github.com/tensorflow/lucid`, accessed 2021-03-17.

[109] Google AI Blog, DeepDream, a Code Example for visualizing neural networks, (2021a), `http://ai.googleblog.com/2015/07/deepdream-code-example-for-visualizing.html`, accessed 2021-03-17.

[110] Google AI Blog, Inceptionism: Going deeper into neural networks, (2021b), `http://ai.googleblog.com/2015/06/inceptionism-going-deeper-into-neural.html`, accessed 2021-03-15.

[111] J. Yosinski, J. Clune, A. Nguyen, T. Fuchs and H. Lipson, Understanding neural networks through deep visualization, (2015), `arXiv:1506.06579 [cs.CV]`.

[112] A. Nguyen, J. Yosinski and J. Clune, Multifaceted feature visualization: Uncovering the different types of features learned by each neuron in deep neural networks, (2016), `arXiv:1602.03616 [cs.NE]`.

[113] A. Nguyen, J. Yosinski and J. Clune, Deep neural networks are easily fooled: High confidence predictions for unrecognizable images, (2015), `arXiv:1412.1897 [cs.CV]`.

[114] A. Nguyen, A. Dosovitskiy, J. Yosinski, T. Brox and J. Clune, Synthesizing

the preferred inputs for neurons in neural networks via deep generator networks, (2016), arXiv:1605.09304 [cs.NE].

[115] C. Olah, A. Mordvintsev and L. Schubert, Feature visualization, *Distill* (2017), doi:10.23915/distill.00007, https://distill.pub/2017/feature-visualization.

[116] C. Olah, A. Satyanarayan, I. Johnson, S. Carter, L. Schubert, K. Ye and A. Mordvintsev, The building blocks of interpretability, *Distill* (2018), doi: 10.23915/distill.00010, https://distill.pub/2018/building-blocks.

[117] A. Mordvintsev, N. Pezzotti, L. Schubert and C. Olah, Differentiable image parameterizations, *Distill* (2018), doi:10.23915/distill.00012, https://distill.pub/2018/differentiable-parameterizations.

[118] Y. LeCun, C. Cortes and C. Burges, Mnist handwritten digit database, *ATT Labs [Online]. Available: http://yann.lecun.com/exdb/mnist* **2** (2010).

[119] M. Ancona, E. Ceolini, C. Öztireli and M. Gross, Towards better understanding of gradient-based attribution methods for deep neural networks, (2018), arXiv:1711.06104 [cs.LG].

[120] M. Alber *et al.*, iNNvestigate neural networks! (2018), arXiv:1808.04260 [cs.LG].

[121] M. Ancona, Marcoancona/DeepExplain, (2021), https://github.com/marcoancona/DeepExplain, accessed 2021-03-26.

[122] Pytorch/captum, (2021), https://github.com/pytorch/captum, accessed 2021-03-26.

[123] B. Zhou, A. Khosla, A. Lapedriza, A. Oliva and A. Torralba, Learning deep features for discriminative localization, (2015), arXiv:1512.04150 [cs.CV].

[124] S. Bach, A. Binder, G. Montavon, F. Klauschen, K. Müller and W. Samek, On pixel-wise explanations for non-linear classifier decisions by layer-wise relevance propagation, *PLoS ONE* **10** (2015).

[125] L. Arras, G. Montavon, K.-R. Müller and W. Samek, Explaining recurrent neural network predictions in sentiment analysis, (2017), arXiv:1706.07206 [cs.CL].

[126] G. Montavon, A. Binder, S. Lapuschkin, W. Samek and K.-R. Müller, *Layer-Wise Relevance Propagation: An Overview*, pp. 193–209. (Springer International Publishing, Cham, 2019), doi:10.1007/978-3-030-28954-6_10, https://doi.org/10.1007/978-3-030-28954-6_10.

[127] L. M. Zintgraf, T. S. Cohen, T. Adel and M. Welling, Visualizing deep neural network decisions: Prediction difference analysis, (2017), arXiv:1702.04595 [cs.CV].

[128] R. C. Fong and A. Vedaldi, Interpretable explanations of black boxes by meaningful perturbation, *2017 IEEE International Conference on Computer Vision (ICCV)* (2017), ISBN 9781538610329, doi:10.1109/iccv.2017.371, http://dx.doi.org/10.1109/ICCV.2017.371.

[129] A. Der Kiureghian and O. Ditlevsen, Aleatory or epistemic? Does it matter? *Structural Safety* **31**, 2, pp. 105–112 (2009).

[130] A. Kendall and Y. Gal, What uncertainties do we need in bayesian deep learning for computer vision? in *NIPS* (2017), arXiv:1703.04977.

[131] B. Nachman, A guide for deploying Deep Learning in LHC searches: How to achieve optimality and account for uncertainty, (2019), arXiv:1909.03081 [hep-ph].

[132] B. Efron, Bootstrap methods: Another look at the jackknife, *The Annals of Statistics* **7**, 1, pp. 1–26 (1979), http://www.jstor.org/stable/2958830.

[133] N. Tagasovska and D. Lopez-Paz, Single-model uncertainties for deep learning, in *NeurIPS* (2019), 1811.00908.

[134] G. Kasieczka, M. Luchmann, F. Otterpohl and T. Plehn, Per-object systematics using deep-learned calibration, *SciPost Phys.* **9**, p. 089 (2020), doi:10.21468/SciPostPhys.9.6.089, arXiv:2003.11099 [hep-ph].

[135] C. Blundell, J. Cornebise, K. Kavukcuoglu and D. Wierstra, Weight uncertainty in neural network, in *International Conference on Machine Learning*. PMLR, pp. 1613–1622 (2015).

[136] S. Chang, T. Cohen and B. Ostdiek, What is the Machine Learning? *Phys. Rev.* **D97**, 5, p. 056009 (2018), doi:10.1103/PhysRevD.97.056009, arXiv:1709.10106 [hep-ph].

[137] L. Bradshaw, R. K. Mishra, A. Mitridate and B. Ostdiek, Mass agnostic jet taggers, (2019), arXiv:1908.08959 [hep-ph].

[138] C. Shimmin *et al.*, Decorrelated jet substructure tagging using adversarial neural networks, *Phys. Rev. D* **96**, 7, p. 074034 (2017), doi:10.1103/PhysRevD.96.074034, arXiv:1703.03507 [hep-ex].

[139] G. Louppe, M. Kagan and K. Cranmer, Learning to pivot with adversarial networks, (2016), arXiv:1611.01046 [stat.ML].

[140] G. Kasieczka and D. Shih, DisCo Fever: Robust networks through distance correlation, (2020), arXiv:2001.05310 [hep-ph].

[141] G. J. Székely and M. L. Rizzo, Brownian distance covariance, *Ann. Appl. Stat.* **3**, 4, pp. 1236–1265 (2009), doi:10.1214/09-AOAS312, https://doi.org/10.1214/09-AOAS312.

[142] G. J. Székely and M. L. Rizzo, The distance correlation t-test of independence in high dimension, *J. Multivar. Anal.* **117**, pp. 193–213 (2013), doi:10.1016/j.jmva.2013.02.012, http://dx.doi.org/10.1016/j.jmva.2013.02.012.

[143] G. J. Székely and M. L. Rizzo, Partial distance correlation with methods for dissimilarities, *Ann. Statist.* **42**, 6, pp. 2382–2412 (2014), doi:10.1214/14-AOS1255, https://doi.org/10.1214/14-AOS1255.

[144] I. J. Goodfellow, J. Shlens and C. Szegedy, Explaining and harnessing adversarial examples, (2014), arXiv:1412.6572 [stat.ML].

[145] B. Nachman and C. Shimmin, AI safety for high energy physics, (2019), arXiv:1910.08606 [hep-ph].

[146] P. De Castro and T. Dorigo, INFERNO: Inference-Aware Neural Optimisation, *Comput. Phys. Commun.* **244**, pp. 170–179 (2019), doi:10.1016/j.cpc.2019.06.007, arXiv:1806.04743 [stat.ML].

[147] S. Eguchi and J. Copas, Interpreting Kullback–Leibler divergence with the Neyman–Pearson lemma, *Journal of Multivariate Analysis* **97**, 9, pp. 2034–2040 (2006), doi:https://doi.org/10.1016/j.jmva.2006.03.007.

[148] J. T. Barron, A more general robust loss function, *CoRR* (2017), arXiv:1701.03077, http://arxiv.org/abs/1701.03077.

[149] S. Greydanus, M. Dzamba and J. Yosinski, Hamiltonian neural networks, *CoRR* (2019), arXiv:1906.01563.

[150] G. E. Hinton, *A Practical Guide to Training Restricted Boltzmann Machines*, pp. 599–619. (Springer, Berlin, Heidelberg, 2012), doi:10.1007/978-3-642-35289-8_32, https://doi.org/10.1007/978-3-642-35289-8_32.

[151] G. Carleo and M. Troyer, Solving the quantum many-body problem with artificial neural networks, *Science* **355**, 6325 (2017), doi:10.1126/science.aag2302, arXiv:1606.02318.

[152] D. Silver *et al.*, Mastering the game of go without human knowledge, *Nature* **550**, 7676, pp. 354–359 (2017).

[153] M. Erdmann, B. Fischer and D. Noll, Reinforced sorting networks for particle physics analyses, *Journal Physics Conference Series* **1525**, p. 012098 (2020), doi:10.1088/1742-6596/1525/1/012098.

[154] J. Brehmer, F. Kling, I. Espejo and K. Cranmer, MadMiner: Machine learning-based inference for particle physics, *Comput. Softw. Big Sci.* **4**, 1, p. 3 (2020), doi:10.1007/s41781-020-0035-2, arXiv:1907.10621 [hep-ph].

[155] Z.-H. Zhou, A brief introduction to weakly supervised learning, *National Science Review* **5**, 1, pp. 44–53 (2018).

[156] B. Settles, Active learning literature survey, Tech. rep., University of Wisconsin-Madison Department of Computer Sciences (2009), http://digital.library.wisc.edu/1793/60660.

[157] J. S. Smith, B. Nebgen, N. Lubbers, O. Isayev and A. E. Roitberg, Less is more: Sampling chemical space with active learning, *The Journal of Chemical Physics* **148**, 24, p. 241733 (2018), doi:10.1063/1.5023802, https://doi.org/10.1063/1.5023802.

[158] O. Chapelle (ed.), *Semi-supervised learning*, Adaptive computation and machine learning. (MIT Press, 2006).

[159] T. G. Dietterich, R. H. Lathrop and T. Lozano-Pérez, Solving the multiple instance problem with axis-parallel rectangles, *Artificial Intelligence* **89**, 1, pp. 31–71 (1997), doi:https://doi.org/10.1016/S0004-3702(96)00034-3.

[160] F. u. A. A. Minhas and A. Ben-Hur, Multiple instance learning of Calmodulin binding sites, *Bioinformatics* **28**, 18, pp. i416–i422 (2012), doi:10.1093/bioinformatics/bts416, https://doi.org/10.1093/bioinformatics/bts416.

[161] S. Bandyopadhyay, D. Ghosh, R. Mitra and Z. Zhao, Mbstar: multiple instance learning for predicting specific functional binding sites in microrna targets, *Scientific Reports* **5**, 1, pp. 1–12 (2015).

[162] G. Quellec, G. Cazuguel, B. Cochener and M. Lamard, Multiple-instance learning for medical image and video analysis, *IEEE Reviews in Biomedical Engineering* **10**, pp. 213–234 (2017), doi:10.1109/RBME.2017.2651164.

[163] J. Amores, Multiple instance classification: Review, taxonomy and comparative study, *Artificial Intelligence* **201**, pp. 81–105 (2013), doi:https://doi.org/10.1016/j.artint.2013.06.003.

[164] P. T. Komiske, E. M. Metodiev, B. Nachman and M. D. Schwartz, Learning to classify from impure samples with high-dimensional data, *Phys. Rev. D* **98**, 1, p. 011502 (2018), doi:10.1103/PhysRevD.98.011502, arXiv:1801.10158 [hep-ph].

[165] B. Frénay and M. Verleysen, Classification in the presence of label noise: A survey, *Neural Networks and Learning Systems, IEEE Transactions on* **25**, pp. 845–869 (2014), doi:10.1109/TNNLS.2013.2292894.

[166] G. Blanchard, M. Flaska, G. Handy, S. Pozzi and C. Scott, Classification with asymmetric label noise: Consistency and maximal denoising, (2016), arXiv:1303.1208 [stat.ML].

[167] E. M. Metodiev, B. Nachman and J. Thaler, Classification without labels: Learning from mixed samples in high energy physics, *JHEP* **10**, p. 174 (2017), doi:10.1007/JHEP10(2017)174, arXiv:1708.02949 [hep-ph].

[168] C. C. Aggarwal, *Neural Networks and Deep Learning.* (Springer International Publishing, 2018).

[169] J. Schmidhuber, Deep learning in neural networks: An overview, *Neural Networks* **61**, pp. 85–117 (2015), arXiv:1404.7828 [cs.NE].

[170] P. Vincent and H. Larochelle, Stacked denoising autoencoders: Learning useful representations in a deep network with a local denoising criterion, *Journal of Machine Learning Research* **11**, pp. 3371–3408 (2010).

[171] T. Heimel, G. Kasieczka, T. Plehn and J. Thompson, QCD or what? *SciPost Physics* **6**, 3 (2019), doi:10.21468/scipostphys.6.3.030, http://dx.doi.org/10.21468/SciPostPhys.6.3.030.

[172] M. Erdmann, F. Schlüter and R. Šmída, Classification and recovery of radio signals from cosmic ray induced air showers with deep learning, *Journal of Instrumentation* **14**, 04, pp. P04005–P04005 (2019), doi:10.1088/1748-0221/14/04/p04005, http://dx.doi.org/10.1088/1748-0221/14/04/P04005.

[173] S. Stille, C. Baeumer, S. Krannich, C. Lenser, R. Dittmann, J. Perlich, S. Roth, R. Waser and U. Klemradt, Feasibility studies for filament detection in resistively switching SrTiO₃ devices by employing grazing incidence small angle X-ray scattering, *Journal of Applied Physics* **113**, p. 064509 (2013), doi:10.1063/1.4792035.

[174] A. Krizhevsky, Learning multiple layers of features from tiny images, (2009), https://www.cs.toronto.edu/~kriz/learning-features-2009-TR.pdf.

[175] C. Fefferman, S. Mitter and H. Narayanan, Testing the manifold hypothesis, *Journal of the American Mathematical Society* **29** (2013), doi:10.1090/jams/852, arXiv:1310.0425.

[176] D. P. Kingma and M. Welling, Auto-encoding variational bayes, (2013), arXiv:1312.6114 [stat.ML].

[177] A. Makhzani, J. Shlens, N. Jaitly and I. J. Goodfellow, Adversarial autoencoders, (2015), arXiv:1511.05644.

[178] I. Higgins, L. Matthey, A. Pal, C. P. Burgess, X. Glorot, M. Botvinick, S. Mohamed and A. Lerchner, Beta-vae: Learning basic visual concepts with a constrained variational framework, in *ICLR* (2017).

[179] R. Ahumada *et al.*, The 16th Data Release of the Sloan Digital Sky Surveys: First Release from the APOGEE-2 Southern Survey and Full Release of eBOSS Spectra, *Astrophys. J. Suppl.* **249**, 1, p. 3 (2020), doi:10.3847/1538-4365/ab929e, arXiv:1912.02905 [astro-ph.GA].

[180] I. J. Goodfellow, J. Pouget-Abadie, M. Mirza, B. Xu, D. Warde-Farley, S. Ozair, A. Courville and Y. Bengio, Generative adversarial networks, (2014), arXiv:1406.2661 [stat.ML].

[181] M. Heusel, H. Ramsauer, T. Unterthiner, B. Nessler, G. Klambauer and S. Hochreiter, GANs trained by a two time-scale update rule converge to a Nash equilibrium, (2017), arXiv:1706.08500.

[182] A. Odena, C. Olah and J. Shlens, Conditional image synthesis with auxiliary classifier gans, (2017), arXiv:1610.09585.

[183] M. Mirza and S. Osindero, Conditional generative adversarial nets, (2014), arXiv:1411.1784.

[184] T. Salimans, I. Goodfellow, W. Zaremba, V. Cheung, A. Radford and X. Chen, Improved techniques for training gans, (2016), arXiv:1606.03498 [cs.LG].

[185] A. Radford, L. Metz and S. Chintala, Unsupervised representation learning with deep convolutional generative adversarial networks, *CoRR* (2016), arXiv:1511.06434 [cs.LG].

[186] M. Arjovsky and L. Bottou, Towards principled methods for training generative adversarial networks, (2017), arXiv:1701.04862.

[187] K. Kurach, M. Lučić, X. Zhai, M. Michalski and S. Gelly, A large-scale study on regularization and normalization in GANs, PMLR, Long Beach, California, USA, pp. 3581–3590 (2019), http://proceedings.mlr.press/v97/kurach19a.html.

[188] A. Odena, V. Dumoulin and C. Olah, Deconvolution and checkerboard artifacts, *Distill* (2016), doi:10.23915/distill.00003, http://distill.pub/2016/deconv-checkerboard.

[189] X. Mao, Q. Li, H. Xie, R. Y. K. Lau, Z. Wang and S. P. Smolley, Least squares generative adversarial networks, *2017 IEEE International Conference on Computer Vision (ICCV)*, pp. 2813–2821 (2017).

[190] M. Arjovsky, S. Chintala and L. Bottou, Wasserstein GAN, (2017), arXiv:1701.07875.

[191] I. Gulrajani, F. Ahmed, M. Arjovsky, V. Dumoulin and A. C. Courville, Improved training of Wasserstein GANs, in *NIPS* (2017), arXiv:1704.00028 [cs.LG].

[192] D. Edwards, On the Kantorovich–Rubinstein theorem, *Expositiones Mathematicae* **29**, 4, pp. 387–398 (2011), doi:https://doi.org/10.1016/j.exmath.2011.06.005, http://www.sciencedirect.com/science/article/pii/S0723086911000430.

[193] J. Ba, J. Kiros and G. E. Hinton, Layer normalization, (2016), arXiv:1607.06450.

[194] T. Miyato, T. Kataoka, M. Koyama and Y. Yoshida, Spectral normalization for generative adversarial networks, (2018), arXiv:1802.05957.

[195] A. Odena, J. Buckman, C. Olsson, T. Brown, C. Olah, C. Raffel and

I. J. Goodfellow, Is generator conditioning causally related to gan performance? (2018), `arXiv:1802.08768`.

[196] A. Brock, J. Donahue and K. Simonyan, Large scale gan training for high fidelity natural image synthesis, (2019), `arXiv:1809.11096`.

[197] Y. Yoshida and T. Miyato, Spectral norm regularization for improving the generalizability of deep learning, (2017), `arXiv:1705.10941`.

[198] G. H. Golub and H. A. van der Vorst, Eigenvalue computation in the 20th century, *Journal of Computational and Applied Mathematics* **123**, 1, pp. 35–65 (2000), doi:https://doi.org/10.1016/S0377-0427(00)00413-1.

[199] A. Odena, Open questions about generative adversarial networks, *Distill* (2019), doi:10.23915/distill.00018, https://distill.pub/2019/gan-open-problems.

[200] X. Wang, R. B. Girshick, A. Gupta and K. He, Non-local neural networks, *2018 IEEE/CVF Conference on Computer Vision and Pattern Recognition*, pp. 7794–7803 (2018).

[201] H. Zhang, I. J. Goodfellow, D. Metaxas and A. Odena, Self-attention generative adversarial networks, (2019), `arXiv:1805.08318`.

[202] T. Karras, S. Laine and T. Aila, A style-based generator architecture for generative adversarial networks, *2019 IEEE/CVF Conference on Computer Vision and Pattern Recognition (CVPR)*, pp. 4396–4405 (2019).

[203] T. Karras, T. Aila, S. Laine and J. Lehtinen, Progressive growing of gans for improved quality, stability, and variation, (2018), `arXiv:1710.10196`.

[204] D. J. Rezende and S. Mohamed, Variational inference with normalizing flows, (2016), `arXiv:1505.05770 [stat.ML]`.

[205] I. Kobyzev, S. Prince and M. Brubaker, Normalizing flows: An introduction and review of current methods, *IEEE Transactions on Pattern Analysis and Machine Intelligence*, p. 1 (2020), doi:10.1109/tpami.2020.2992934, `http://dx.doi.org/10.1109/TPAMI.2020.2992934`.

[206] J. Brehmer and K. Cranmer, Flows for simultaneous manifold learning and density estimation, (2020), `arXiv:2003.13913 [stat.ML]`.

[207] G. Papamakarios, E. Nalisnick, D. J. Rezende, S. Mohamed and B. Lakshminarayanan, Normalizing flows for probabilistic modeling and inference, (2019), `arXiv:1912.02762 [stat.ML]`.

[208] L. Dinh, J. Sohl-Dickstein and S. Bengio, Density estimation using real nvp, (2017), `arXiv:1605.08803 [cs.LG]`.

[209] S. T. Radev, U. K. Mertens, A. Voss, L. Ardizzone and U. Köthe, Bayesflow: Learning complex stochastic models with invertible neural networks, (2020), `arXiv:2003.06281 [stat.ML]`.

[210] G. Papamakarios, T. Pavlakou and I. Murray, Masked autoregressive flow for density estimation, (2018), `arXiv:1705.07057 [stat.ML]`.

[211] M. S. Albergo, D. Boyda, D. C. Hackett, G. Kanwar, K. Cranmer, S. Racanière, D. J. Rezende and P. E. Shanahan, Introduction to normalizing flows for lattice field theory, (2021), `arXiv:2101.08176 [hep-lat]`.

[212] G. Kanwar, M. S. Albergo, D. Boyda, K. Cranmer, D. C. Hackett, S. Racanière, D. J. Rezende and P. E. Shanahan, Equivariant flow-based sampling for lattice gauge theory, *Physical Review Letters* **125**, 12

(2020), doi:10.1103/physrevlett.125.121601, http://dx.doi.org/10.1103/PhysRevLett.125.121601.

[213] C. Gao, S. Höche, J. Isaacson, C. Krause and H. Schulz, Event Generation with Normalizing Flows, *Physics Review D* **101**, 7, p. 076002 (2020), doi: 10.1103/PhysRevD.101.076002, arXiv:2001.10028 [hep-ph].

[214] B. Nachman and D. Shih, Anomaly Detection with Density Estimation, *Phys. Rev. D* **101**, p. 075042 (2020), doi:10.1103/PhysRevD.101.075042, arXiv:2001.04990 [hep-ph].

[215] T. Glüsenkamp, Unifying supervised learning and vaes — automating statistical inference in high-energy physics, (2020), arXiv:2008.05825 [cs.LG].

[216] S. Barratt and R. Sharma, A note on the inception score, (2018), arXiv:1801.01973.

[217] A. Butter, S. Diefenbacher, G. Kasieczka, B. Nachman and T. Plehn, Ganplifying event samples, (2020), arXiv:2008.06545 [hep-ph].

[218] T. Quast, *Qualification, performance validation and fast generative modelling of beam test calorimeter prototypes for the CMS calorimeter endcap upgrade*, Dissertation, RWTH Aachen University, Aachen (2020), doi:10.18154/RWTH-2020-06473, https://publications.rwth-aachen.de/record/792901.

[219] M. Erdmann, J. Glombitza and T. Quast, Precise simulation of electromagnetic calorimeter showers using a wasserstein generative adversarial network, *Computing and Software for Big Science* **3**, pp. 1–13 (2019), doi: 10.1007/s41781-018-0019-7.

[220] M. Paganini, L. de Oliveira and B. Nachman, Calogan: Simulating 3d high energy particle showers in multi-layer electromagnetic calorimeters with generative adversarial networks, (2017), arXiv:1712.10321.

[221] V. Chekalina, E. Orlova, F. Ratnikov, D. Ulyanov, A. Ustyuzhanin and E. Zakharov, Generative models for fast calorimeter simulation.lhcb case, (2018), arXiv:1812.01319.

[222] P. Musella and F. Pandolfi, Fast and accurate simulation of particle detectors using generative adversarial networks, *Computing and Software for Big Science* **2**, 1 (2018), doi:10.1007/s41781-018-0015-y, http://dx.doi.org/10.1007/s41781-018-0015-y.

[223] C. Ahdida *et al.*, Fast simulation of muons produced at the SHiP experiment using generative adversarial networks, *Journal of Instrumentation* **14**, 11, pp. P11028–P11028 (2019), doi:10.1088/1748-0221/14/11/p11028, https://doi.org/10.1088/1748-0221/14/11/p11028.

[224] D. Belayneh *et al.*, Calorimetry with deep learning: Particle simulation and reconstruction for collider physics, (2019), arXiv:1912.06794.

[225] R. Kansal *et al.*, Graph generative adversarial networks for sparse data generation in high energy physics, (2021), arXiv:2012.00173 [physics.data-an].

[226] A. C. Rodríguez *et al.*, Fast cosmic web simulations with generative adversarial networks, *Computational Astrophysics and Cosmology* **5**, 1 (2018), doi:10.1186/s40668-018-0026-4, http://dx.doi.org/10.1186/s40668-018-0026-4.

[227] M. Erdmann, L. Geiger, J. Glombitza and D. Schmidt, Generating and Refining Particle Detector Simulations using the Wasserstein Distance in Adversarial Networks, *Computing and Software for Big Science* **2**, pp. 1–9 (2018), arXiv:1802.03325 [astro-ph.IM].

[228] A. Martelli, The CMS HGCAL detector for HL-LHC upgrade, (2017), arXiv:1708.08234 [physics.ins-det].

[229] S. Agostinelliae *et al.*, Geant4 — a simulation toolkit, *Nuclear Instruments & Methods in Physics Research Section A* **506**, pp. 250–303 (2018).

[230] J. Halverson and C. Long, Statistical predictions in string theory and deep generative models, *Fortschritte der Physik* **68**, 5, p. 2000005 (2020), doi: 10.1002/prop.202000005, http://dx.doi.org/10.1002/prop.202000005.

[231] K. Weiss, T. M. Khoshgoftaar and D. Wang, A survey of transfer learning, *Journal of Big data* **3**, 1, pp. 1–40 (2016).

[232] A. Andreassen and B. Nachman, Neural networks for full phase-space reweighting and parameter tuning, *Physics Review D* **101**, 9, p. 091901 (2020), doi:10.1103/PhysRevD.101.091901, arXiv:1907.08209 [hep-ph].

[233] The CMS Collaboration, A. M. Sirunyan *et al.*, Identification of heavy, energetic, hadronically decaying particles using machine-learning techniques, *JINST* **15**, 06, p. P06005 (2020), doi:10.1088/1748-0221/15/06/P06005, arXiv:2004.08262 [hep-ex].

[234] M. Erdmann, B. Fischer, D. Noll, Y. Rath, M. Rieger and D. Schmidt, Adversarial Neural Network-based data-simulation corrections for jet-tagging at CMS, *Journal Physics Conference Series* **1525**, p. 012094 (2020), doi: 10.1088/1742-6596/1525/1/012094.

[235] Y. Ganin *et al.*, Domain-adversarial training of neural networks, *Journal of Machine Learning Research* **17**, 59, pp. 1–35 (2016), arXiv:1505.07818 [stat.ML].

[236] The CMS Collaboration, A. M. Sirunyan *et al.*, A deep neural network to search for new long-lived particles decaying to jets, *Machine Learning Science Technology* **1**, p. 035012 (2020), doi:10.1088/2632-2153/ab9023, arXiv:1912.12238 [hep-ex].

[237] M. Bellagente *et al.*, Invertible networks or partons to detector and back again, *SciPost Physics* **9**, p. 074 (2020), doi:10.21468/SciPostPhys.9.5.074, arXiv:2006.06685 [hep-ph].

[238] A. Gretton, K. Borgwardt, M. Rasch, B. Schölkopf and A. Smola, A kernel method for the two-sample-problem, *Advances in Neural Information Processing Systems* **19**, pp. 513–520 (2006).

[239] A. Andreassen, P. T. Komiske, E. M. Metodiev, B. Nachman and J. Thaler, OmniFold: A method to simultaneously unfold all observables, *Physics Review Letter* **124**, 18, p. 182001 (2020), doi:10.1103/PhysRevLett.124.182001, arXiv:1911.09107 [hep-ph].

[240] S. Diefenbacher, E. Eren, G. Kasieczka, A. Korol, B. Nachman and D. Shih, DCTRGAN: Improving the Precision of Generative Models with Reweighting, *JINST* **15**, 11, p. P11004 (2020), doi:10.1088/1748-0221/15/11/P11004, arXiv:2009.03796 [hep-ph].

[241] J.-Y. Zhu, T. Park, P. Isola and A. A. Efros, Unpaired image-to-image

translation using cycle-consistent adversarial networks, *2017 IEEE International Conference on Computer Vision (ICCV)*, pp. 2242–2251 (2017).

[242] G. Kasieczka *et al.*, The LHC Olympics 2020: A community challenge for anomaly detection in high energy physics, (2021), arXiv:2101.08320 [hep-ph].

[243] The CMS Collaboration, A. M. Sirunyan *et al.*, MUSiC: a model unspecific search for new physics in proton-proton collisions at $\sqrt{s} = 13$ TeV, (2020), arXiv:2010.02984 [hep-ex].

[244] The ATLAS Collaboration, M. Aaboud *et al.*, A strategy for a general search for new phenomena using data-derived signal regions and its application within the atlas experiment, *The European Physical Journal C* **79**, 2 (2019), doi:10.1140/epjc/s10052-019-6540-y, http://dx.doi.org/10.1140/epjc/s10052-019-6540-y.

[245] G. E. Hinton and R. R. Salakhutdinov, Reducing the dimensionality of data with neural networks, *Science* **313**, 5786, pp. 504–507 (2006), doi:10.1126/science.1127647.

[246] S. Hawkins, H. He, G. Williams and R. Baxter, Outlier detection using replicator neural networks, in *International Conference on Data Warehousing and Knowledge Discovery*, pp. 170–180. (Springer, 2002).

[247] L. Ruff *et al.*, Deep one-class classification, in J. Dy and A. Krause (eds.), *Proc. 35th Int. Conf. on Machine Learning*, Vol. 80. PMLR, Stockholm Sweden, pp. 4393–4402 (2018), http://proceedings.mlr.press/v80/ruff18a.html.

[248] J. H. Collins, K. Howe and B. Nachman, Anomaly detection for resonant new physics with machine learning, *Physics Review Letters* **121**, 24, p. 241803 (2018), doi:10.1103/PhysRevLett.121.241803, arXiv:1805.02664 [hep-ph].

[249] The Institute for Ethical AI & Machine Learning, Awesome production machine learning, (2021), https://github.com/EthicalML/awesome-production-machine-learning.

[250] H. Qin, R. Gong, X. Liu, X. Bai, J. Song and N. Sebe, Binary neural networks: A survey, *Pattern Recognition* **105**, p. 107281 (2020), doi:10.1016/j.patcog.2020.107281, http://dx.doi.org/10.1016/j.patcog.2020.107281.

[251] D. Modha, Introducing a brain-inspired computer, (2017), https://www.research.ibm.com/articles/brain-chip.shtml.

[252] A. Marban, D. Becking, S. Wiedemann and W. Samek, Learning sparse & ternary neural networks with entropy-constrained trained ternarization (ec2t), (2020), arXiv:2004.01077 [cs.LG].

[253] F. Fahim *et al.*, hls4ml: An open-source codesign workflow to empower scientific low-power machine learning devices, (2021), arXiv:2103.05579 [cs.LG].

[254] M. Schuld, I. Sinayskiy and F. Petruccione, An introduction to quantum machine learning, *Contemporary Physics* **56**, 2 (2014), doi:10.1080/00107514.2014.964942, arXiv:1409.3097.

[255] J. Biamonte, P. Wittek, N. Pancotti, P. Rebentrost, N. Wiebe and

S. Lloyd, Quantum machine learning, *Nature* **549**, 7671 (2017), doi: 10.1038/nature23474, arXiv:1611.09347.

[256] D. Marković, A. Mizrahi, D. Querlioz and J. Grollier, Physics for neuromorphic computing, *Nature Reviews Physics* **2**, 9, pp. 499–510 (2020), doi:10.1038/s42254-020-0208-2, arXiv:2003.04711 [cs.ET].

[257] S. Furber *et al.*, Spinnaker (spiking neural network architecture), (2021), http://apt.cs.manchester.ac.uk/projects/SpiNNaker.

[258] K. Meier *et al.*, Silicon brains, (2018), https://www.humanbrainproject.eu/en/silicon-brains.

[259] A. Mehonic, A. Sebastian, B. Rajendran, O. Simeone, E. Vasilaki and A. J. Kenyon, Memristors — from in-memory computing, deep learning acceleration, spiking neural networks, to the future of neuromorphic and bio-inspired computing, (2020), arXiv:2004.14942 [cs.ET].

[260] J. Sacramento, R. P. Costa, Y. Bengio and W. Senn, Dendritic cortical microcircuits approximate the backpropagation algorithm, *Advances in Neural Information Processing Systems 31 (NIPS 2018)* (2018), arXiv:1810.11393.

[261] G. Bellec, F. Scherr, A. Subramoney, E. Hajek, D. Salaj, R. Legenstein and W. Maass, A solution to the learning dilemma for recurrent networks of spiking neurons, *bioRxiv* (2019), doi:10.1101/738385, https://www.biorxiv.org/content/early/2019/08/31/738385.

[262] A. Tavanaei, M. Ghodrati, S. R. Kheradpisheh, T. Masquelier and A. Maida, Deep learning in spiking neural networks, *Neural Networks* **111**, pp. 47–63 (2019), doi:10.1016/j.neunet.2018.12.002, http://dx.doi.org/10.1016/j.neunet.2018.12.002.

[263] T. A. Enßlin, Information theory for fields, *Annalen Phys.* **531**, 3, p. 1800127 (2019), doi:10.1002/andp.201800127, arXiv:1804.03350 [astro-ph.CO].

[264] M. Cranmer *et al.*, Discovering symbolic models from deep learning with inductive biases, (2020), arXiv:2006.11287 [cs.LG].

Index